RADIATION PROTECTION

ICRP PUBLICATION 18

The RBE for High-LET Radiations with respect to Mutagenesis

A Report prepared by a task group of Committee 1 of the International Commission on Radiological Protection

Adopted by the Commission in May 1972

PUBLISHED FOR

The International Commission on Radiological Protection

Pergamon Press Ltd., Headington Hill Hall, Oxford

Pergamon Press Inc., Maxwell House, Fairview Park, Elmsford,
New York 10523

Pergamon of Canada Ltd., 207 Queen's Quay West, Toronto 1

Pergamon Press (Aust.) Pty. Ltd., 19a Boundary Street,
Rushcutters Bay, N.S.W. 2011, Australia

Vieweg & Sohn GmbH, Burgplatz 1, Braunschweig

First edition 1972

Library of Congress Catalog Card No. 72-86898

This is one of a series of reports prepared as background material for the International Commission on Radiological Protection. These reports, published in blue covers, form part of the Commission's continuing review of information intended to provide scientific bases for its Recommendations, which are published in brown covers. The Commission hopes that the publication of the reports in blue covers, while not necessarily implying recommendations for present action, will stimulate discussion on matters having direct relevance to its work and to the development of the fundamental principles of radiological protection.

ISBN 0 08 017008 0

Contents

Preface

AT its meeting in Oxford in 1969, the International Commission on Radiological Protection (ICRP) approved the formation by Committee 1 of a Task Group on the RBE for Neutrons with respect to Mutagenesis, with membership as follows:

A. G. SEARLE (*Chairman*)
F. J. DE SERRES
G. J. NEARY
W. L. RUSSELL
H. H. SMITH

Corresponding members were W. F. BALDWIN, B. A. BRIDGES, A. MURAKAMI, Y. NAKAO and M. D. POMERANTZEVA. A. L. BATCHELOR and J. R. K. SAVAGE attended one meeting and their help is gratefully acknowledged. The terms of reference of the Task Group were "To report on the substantial data on the mutagenic effectiveness of neutrons and other high-LET radiations that have accumulated since the RBE Report, and to point out any implications of these new data for radiation protection." The report is presented herewith. Since the Task Group considered relevant studies on the effects of any radiation of high LET, its title was later amended by the substitution of "high-LET radiations" for "neutrons".

The RBE for High-LET Radiations with respect to Mutagenesis

Introduction

It has been known for many years that radiations of high-LET (linear energy transfer) tend to be more effective for the induction of somatic radiation damage than low, but it is only quite recently that the mutagenic effectiveness of such radiations has been investigated in detail. In its 1963 report to the International Commissions on Radiological Protection and on Radiological Units and Measurements, the RBE Committee considered the limited amount of information then available on the genetic effects of high-LET radiations, which mainly stemmed from work on the fruit fly *Drosophila*. It decided that RBEs* for fast neutrons relative to x and γ rays were probably not much above unity for the induction of recessive lethals and visibles at high dose rates, but some increase in RBE might be anticipated at lower doses and/or dose rates. For spermatozoal dominant lethal induction, however, RBE values of 4–6 were found at high doses and dose rates. Thus there seemed to be some differences between RBEs for gross chromosomal mutations and for point mutations. The importance of dose rate was underlined by results of work on chromosome aberrations in plants, in which very high maximum values (up to 100) of the RBE (RBE_M) were obtained for both fission and 3 MeV neutrons by extrapolation to low doses and dose rates.

The RBE Committee concluded that: "For genetic effects there are indications that RBE values under the exposure conditions envisaged in protection recommendations may be lower than for somatic effects, but it would seem wiser to wait until experimental RBE data are available regarding the production of visible mutations at low levels of dose and dose rate before recommending any change in procedure." Now that a large volume of experimental work on mutagenic effects of high-LET radiations has been carried out on a wide variety of organisms, the time is clearly ripe for a reassessment of RBE values for genetic effects and their implications for protection.

Various reports of international bodies, issued since the RBE Report, have discussed the genetic effects of high-LET radiation, but not in great detail. The ICRP Task Group on the Biological Effects of High-energy Radiations (1966) considered radiobiological aspects of supersonic transport, including genetic aspects. It pointed out that preliminary figures for the yield of mutations from protracted exposures of male mice to fast neutrons suggested that neutron gamma RBEs for spermatogonial irradiation were higher than RBEs for genetic effects considered by the RBE Committee (1963). As an interim measure, however, it recommended no change in the Quality Factor values given in the RBE Committee's Report and adopted by the ICRP (1966).

The 1966 UNSCEAR Report concluded that RBE values for genetic effects of neutrons were almost always in the range of 1 to 6 in experimental animals, though higher values had been obtained at low dose rates. The dependence of the RBE on dose, dose rate, germ-cell stage and type of genetic damage scored was emphasized. With respect to human population exposure,

---

* The meanings of RBE (relative biological effectiveness) were discussed by the RBE Committee. If A rads of a high-LET radiation and B rads of a low-LET radiation (usually x rays) are needed to produce a given experimental effect, then the RBE of the high-LET radiation is B/A. In genetic work it may be expressed as the ratio of the mutation rates induced by the high- and the low-LET radiation, under similar conditions. It is important to note that each RBE value refers to a particular type of effect and that values depend on many factors (e.g. dose and dose rate).

1

the Report states: "It seems that in spermatogonia the rate of induction of mutations per
unit dose of neutrons may be some twenty times
higher than the corresponding rate for x or
γ rays." The 1969 UNSCEAR Report on radiation-induced chromosome aberrations in human
cells summarized data on neutron x ray RBEs
for induction of chromosomal damage in human
peripheral lymphocytes. The few values quoted
lie between 2 and 5; these were for acute exposures.

Scope of Report

Direct evidence from man on the mutagenic
effectiveness of high-LET radiations is very
limited indeed. Another mammal, the laboratory
mouse, has been studied in some depth, but it
would be unwise to rely entirely on information
from a single experimental species in attempting
to assess the probable genetic effects of high-LET
radiations in man. This is especially true under
the present circumstances because: (i) rodents
and primates are distantly related mammals,
and comparative studies on mutagenesis have
not yet bridged the gap between them; (ii) work
on high-LET radiation mutagenesis in the mouse
has been mainly concerned up to now with the
effects of fission neutrons, and so cannot give
much information on the general relationship
between RBE and LET with respect to mutagenesis.

For these and other reasons the Task Group
felt it advisable to attempt a wide-ranging comparative survey of work on high-LET mutagenesis, including information from micro-organisms,
fungi, higher plants and insects as well as mammals. However, only those studies with a direct
bearing on the problems of genetic effects have
been included. Thus, most work on radiation
effects in somatic cells has been omitted, apart
from (i) data on genetic effects in somatic cell
lineages, which have thrown light on the RBE–
LET relationship, and (ii) some findings on the
nduction of chromosomal aberrations in human
ymphocytes by fast neutron irradiation, which

supplement the very meagre amount of direct
information on genetic effects in man. Some
fundamental studies on the mechanisms of radiation mutagenesis which help to throw light on
the relative effectiveness of high-LET radiations
and on the reasons for variation in mutagenic
response have also been included.

The Task Group was asked to point out the
implications of recent data on high-LET mutagenesis from the point of view of radiation protection. We have considered this problem with
particular reference to the Commission's present
recommendations given in ICRP Publication 9
(1966). This tabulates a series of Quality Factors
(QFs)* applicable to different LET values. The
Commission recommended that the absorbed
dose in rads be multiplied by the appropriate
QF in order to obtain a Dose Equivalent in
rems which can be used for radiation protection
calculations. The Group has felt it appropriate
to give an opinion on whether the published
QFs require any modification with respect to
genetic effects in the light of recent findings.

This report includes information on the effects
of high and low doses administered at high and
low dose rates. The relevance of these to radiation protection will be discussed. The distinction
between high- and low-LET radiation is also
discussed later.

Present and Potential Human Exposures to High-LET Radiation

The average dose to human gonads derived
from high-LET radiation in the form of α-parti-
cles emitted by internally-deposited radio-iso-
topes (mainly ^{210}Po and ^{222}Rn) is about 0.6
mrad/yr (United Nations, 1966) or 6 mrem/yr
if the RBE is taken as 10. Other sources of exposure to high-LET radiation in peacetime come
under the headings of (i) cosmic radiation and
(ii) artificial sources used for generating power
or for radiation therapy.

---

* The abbreviation "Q" has been recommended by
ICRU (1971).

Cosmic Radiation

The dose rate from this source increases with altitude; the dose per flight in supersonic transport as compared with present subsonic aircraft depends on the relative durations of the journeys. The information given here comes mainly from an ICRP Task Group Report (1966) on the radiobiological effects of such transport. The components of cosmic radiation are:

(a) *Galactic radiation*, consisting primarily of energetic protons, α particles and some heavier nuclei (up to Z values of about 50). These impinge continually on the earth's atmosphere. Their intensity tends to show a progressive increase with altitude (Fig. 1), but varies with latitude and solar activity. Dose rates seldom exceed 1 mrad/hr.

Neutrons are produced as secondaries by the cosmic rays. Their dose rate also depends on both latitude and altitude, but chiefly on the latter. Figure 1 illustrates the relationship between dose rates of various kinds of radiation and altitude at a geomagnetic latitude of 55°. Schaefer (1968, 1969) has discussed probable levels of exposure at supersonic transport altitudes and estimated a neutron dose-equivalent rate at 20 km of 0.49 mrem/hr, using a quality factor of 8, compared to an overall mean rate of about 1 mrem/hr at this altitude. O'Brien and McLaughlin (1970) calculated that if an aircraft crew member flew at 20 km for about 500 hr/yr the total annual dose equivalent from galactic cosmic rays might exceed 600 mrem.

Even at high altitudes heavy primaries will have mainly disintegrated into smaller fragments.

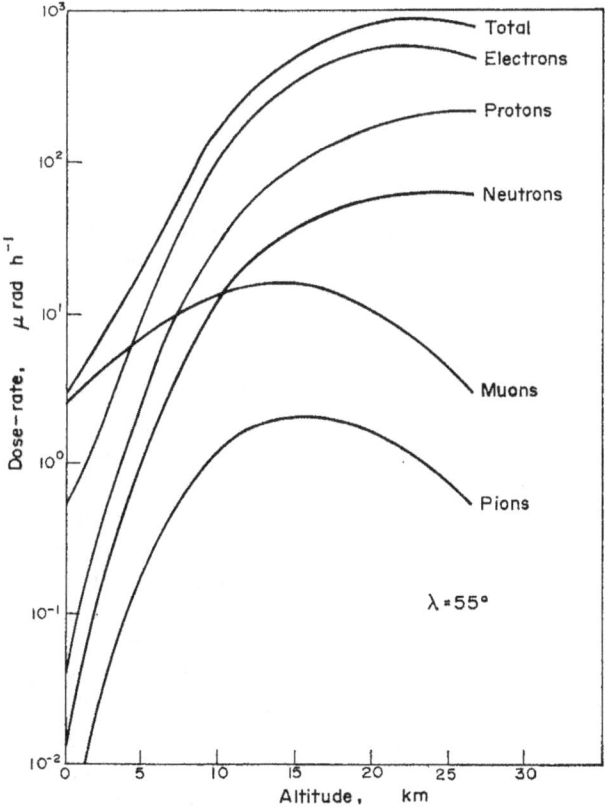

Fig. 1. Variation of galactic radiation dose rates with altitude at geomagnetic latitude of 55°. From O'Brien and McLaughlin (1970).

Besides the neutrons dealt with above there will be secondary protons and α particles of high LET, but the contributions of these components to the total dose is likely to be small.

(b) *Solar radiation*, of which the most important component is that associated with solar flares. Beams of solar protons enter the atmosphere continually, but the bulk of these have limited penetrating power. Neutrons may contribute a large part of the dose at the altitudes used by supersonic transport.

Large solar flares may include particles with energies up to and exceeding 1 GeV, which may give a dose of 1–2 rad during the first hour of such an event at an altitude of 18 000–24 000 m in the polar region. Neutrons of 0.1–10 MeV contribute the bulk of the neutron dose equivalent with an overall effective QF (derived from LET values) of 8.5. The neutron dose is up to 30% of the total. The average solar radiation at 18 000 m was estimated to be 0.045 mrad/hr with a QF of 3.3 giving 0.15 mrem/hr.

ARTIFICIAL SOURCES

(a) *Nuclear reactors and ancillary processes*

Adequate shielding and extensive use of monitoring equipment have kept personnel exposures low. Some figures for the United Kingdom illustrate this. In March 1970, the total annual dose received in the United Kingdom from those covered by the fast neutron film service of the Radiological Protection Service (RPS)* was estimated to be roughly 400 man-rem (Dunster, 1970). The significant exposures were probably received by between 1 000 and 2 000 people, with the maximum dose very rarely exceeding 2 rem/yr. The type of film and method of dosimetry are described by Stevenson (1964). There were separate monitoring schemes for personnel of the Atomic Energy Research Establishment, Harwell (AERE), and the Central Electricity Generating Board (CEGB). Details of the dosimetric methods used in AERE are

* The functions of the RPS have now been taken over by the National Radiological Protection Board, Harwell, Didcot, Berks., U.K.

given by Cook (1958). In 1970, the total recorded dose equivalent on the neutron track plates was only 32 man-rem, 25 of which can be attributed to background α-tracks. About 600 personnel received significant exposures (Smith, J. W., 1971). The CEGB figures for 1970 (Hill, 1971) show that only two out of 954 personnel wearing fast neutron film dosimeters received a dose-equivalent of more than 0.03 rem, while 200 out of 1 230 personnel with "albedo" dosimeters (which detect a range of neutron energies from thermal to fast) received more than 0.03 rem. However, only one of them received more than 0.1 rem. Rather higher doses are received during manufacture of fast neutron (α, *n*) sources, with energies up to 10–14 MeV, at the Radiochemical Centre (Dancer, 1971). The annual neutron dose-equivalent is about 10 man-rem among 20–30 people. Four of these received more than 1 rem from October 1970 to September 1971, with a maximum of 1.6 rem. Exposure levels are expected to fall in a few years' time with improved facilities.

In the Netherlands in 1970, 969 personnel wore fast neutron film badges (described by van der Feer *et al.*, 1964) and received a total accumulated fast neutron dose-equivalent of about 3 man-rem (van der Feer, 1971).

(b) *Therapeutic use of neutrons*

This is also covered in the United Kingdom by the neutron film service of the RPS. The present contribution to the total dose is very small (Dunster, 1970). Over the next decade some expansion in the therapeutic uses of neutrons seems probable, with a consequent slight additional exposure of more hospital personnel.

In conclusion, population exposure to high-LET radiations in the United Kingdom seems to be well below present-day dose limits. In the future it may rise slightly because of (i) increased travel in supersonic aircraft, (ii) expanding use of nuclear reactors for power, (iii) more employment of neutrons for therapy. Although it has not yet been possible to obtain much comparable data from other countries, it seems unlikely that high-LET components of occupational exposures received by any group of atomic energy workers outside the United Kingdom will be much higher than those within the U.K.

Types and Characteristics of the High-LET Radiations used in Mutagenesis Research

TYPES OF RADIATION

The term "high-LET radiations" and its implications call for some comment. A given quality of radiation is not assigned to the low- or high-LET class merely by reference to some arbitrarily chosen numerical boundary on the LET scale. A basis for a meaningful classification is provided by the fact that there is generally a progressive qualitative change in the effects as the LET is increased. Rossi (1971) has recently tabulated a number of such features which are different in most systems in the two LET regions below 10 keV/μm and above 100 keV/μm; these are shown in Table 1. While Rossi would prefer to restrict the term "high-LET" to the region above 100 keV/μm, he points out that for several of the radiations met with in practice (e.g. fast neutrons) most of the LET distribution may fall between the arbitrary limits of 10 keV/μm and 100 keV/μm, but that the biological effects tend to those characteristic of his "high-LET" category.

No hard and fast universal simple criterion is available, but it seems safe to say that in respect of biological effectiveness, dependence of radio-sensitivity on oxygen, survival curve shape and recovery, 50 keV/μm would generally be classed as a high-LET.

A wide variety of high-LET radiations has been used in studies of radiation mutagenesis. In many cases, the choice of the type of radiation was determined more by what was locally available rather than by considerations of what would be ideally desirable in a coordinated plan to obtain with maximum economy empirical data for practical use in hazard evaluation on the one hand and for elucidation of the mechanisms of radiation mutagenesis on the other. A summary of the radiations used is given in Table 2.

RADIATION QUALITY

Some means of comparison of different qualities of radiation is required. If it were possible, a characterization of a particular radiation quality in terms of the numerical value of a common index of quality would be desirable. The term *quality* has been well explained in ICRU Report 16 (1970): the term "refers to those features of the *spatial* distribution of energy transfers—along and within the tracks of particles—that influence the effectiveness of an irradiation in producing change,

TABLE 1
ACTION OF HIGH- AND LOW-LET RADIATION ON MAMMALIAN CELLS IN TISSUE CULTURE
(After Rossi, 1971)

	Low-LET (10 keV/μm)	High-LET (100 keV/μm)
Cell survival curve (at low doses)	Sigmoid	Exponential
Sublethal damage	Present	Absent
Recovery*	Present	Absent
Biological effectiveness	Low with little dependence of LET	High but decreasing with increasing LET
Change of sensitivity during cell division cycle	Pronounced	Minor
Effect of dose modifiers	Appreciable	Negligible
Oxygen enhancement ratio	2.5–3.0	1
Dependence of radio-sensitivity on cell type	Considerable and probably complex	Minor and probably proportional to nuclear diameter

* This term applies to intracellular recovery of the Elkind type and not to tissue repair processes.

TABLE 2
TYPES OF RADIATION USED

Thermal neutrons	
Intermediate neutrons	
Fast neutrons	(i) Monoenergetic neutrons
	(ii) "Cyclotron" neutrons
	(iii) Fission neutrons
	(iv) Polonium–beryllium neutrons
Charged particles	(i) Alpha particles
	(ii) Accelerated particles, ions of hydrogen (protons), deuterium, helium, lithium, boron, carbon, oxygen, neon, argon, at energies up to about 10 MeV per atomic mass unit
	(iii) Very fast protons and helium ions, at energies of several hundred MeV
	(iv) Negative π mesons, producing protons, deuterons, tritons and α particles by nuclear interaction

when other physical factors such as total energy dissipated, absorbed dose, absorbed dose rate and absorbed dose fractionation are kept constant". ICRU Report 16 goes on to explain why the spatial distribution of energy transfers can influence the biological effectiveness of the radiation, depending on the mechanism of production of the particular effect in question, and why the most generally useful simple index of radiation quality is the linear energy transfer (LET).

LET

The Report of the RBE Committee (1963) and ICRU Report 16 (1970) discuss the concept of LET, its complexities and its limitations. The energy lost by an ionizing particle in traversing a small segment of its track is not all communicated to the medium in the form of ionizations and excitations precisely along the axis of the track. The more energetic collisions of the primary particle produce fast secondary electrons i.e. δ rays, which can travel appreciable distances from the primary track before dissipating all their energy. In considering detailed mechanisms of radiation action, this aspect of LET has to be

taken into account, but for purposes of radiological protection the fact that ionization events are distributed in a finite region around a track is usually ignored and the radiation quality is measured in terms of the LET_∞, that is, the energy lost by the particle per unit length of track segment regardless of the actual distribution of the energy given up to the medium surrounding the track segment (ICRP Publication 9, 1966, p. 3).

The LET_∞ varies along the track of a particle; it increases as the particle slows down, reaches a peak shortly before the particle stops, and rapidly falls to zero at the end of the track (cf. the Bragg curve for the variation of ionization produced in a gas along the track of a particle). The minimum LET_∞ of a fast singly-charged particle such as an electron or proton is about 0.2 keV/μm in tissue or water; when the particle has slowed down to bring it into the region of high-LET, say some tens of keV/μm, the remaining track length is quite short, commonly less than 100 μm. Thus, irradiation at a *single* high value of LET_∞ is only practicable when a monolayer of cells is irradiated by charged particles with total track length considerably greater than the cell diameter (Neary, 1970), as in some of the microbiological work reviewed in this report. Even then, there is in essence a distribution of radiation quality which depends on track structure and the existence of δ rays. Some distributions of LET with various arbitrary cut-off values for the energy transfers which are considered to be dissipated locally at the track are illustrated in ICRU Report 16 (1970), in Howard-Flanders (1958), in Fluke et al. (1960), and in Brustad (1962) for various charged particles with energies around 10 MeV per atomic mass unit which have often been used. Agreement between various published data is imperfect because of uncertainty about the LET of low-energy electrons.

LET DISTRIBUTIONS

In most situations of radiation protection and in many of those of radiobiological experimentation there is unavoidably a distribution of radiation quality because the size of the irradiated specimen is much larger than the range of a high-LET particle. Frequently, the charged particles are generated in the irradiated tissue by the interaction of indirectly-ionizing radiation such as x or δ rays or neutrons. Thus any point in the medium is liable to be traversed by any part of the track of a charged particle and so the LET at the point may assume any value between the minimum value at the beginning of the track and the maximum at the Bragg peak near the end of the track.

In addition, the interaction of an x or γ quantum, or of a neutron having a particular energy, with an absorbing medium does not normally lead to a secondary charged particle with a unique initial energy; the initial energy may be anywhere between some maximum value and zero. Moreover, the primary radiation itself frequently has a distribution of energies. Common examples are 250 kV x rays and fission neutrons; data on such energy distributions for neutrons are given in ICRU Report 13 (1969).

Examples of LET distributions arising from the combination of some or all of these sources of heterogeneity are given in Boag (1954),

Bruce et al. (1963), Bewley (1968), and ICRU Report 16 (1970). In some cases, the distribution of LET_∞ is given; in others details of track structure have been taken into account and the distribution refers to the restricted LET corresponding to some energy cut-off.

When evaluating the LET distribution in the irradiation of an animal or any other object with a finite size, it is also necessary to consider the fact that the energy spectrum of the incident radiation may be changed as the radiation penetrates into the specimen. Data for fission neutrons in small-animal sized cylinders are given by Wilhoit and Jones (1970); related data are given by Auxier et al. (1969) and by Frigerio and Branson (1969).

AVERAGE LET

Although a complete LET distribution for a particular irradiation situation contains a great deal of the basic physical information required to characterize that situation, simple intercomparison of different irradiation situations only becomes practicable if some single parameter of radiation quality can be used. Some kind of average LET value would appear a reasonable choice, but a difficulty arises from uncertainty as to the most appropriate kind of average. Without knowledge of the mechanism of production of a biological effect, no clear decision can be made. Two particular kinds of average have usually been considered; for dose average, $\bar{L}_D$, the weighting factor for the different components of the LET distribution is the fractional dose due to any component, while for the track average, $\bar{L}_T$, the weighting factor is the fraction of the total track length associated with any component (RBE Report 1963; ICRU Report 16, 1970). When the LET distribution is a wide one, the numerical values of the two kinds of average are quite different; for example, for hard x rays, $\bar{L}_D$ is of the order of 10 keV/μm and $\bar{L}_T$ of the order of 1 keV/μm (Bruce et al., 1963). Bewley (1968) has given values for various fast neutron spectra and has pointed out that the track average is correlated inversely with mean neutron energy whereas the dose average varies little with neutron energy.

Randolph (1964) showed that dose-average LET would be appropriate if the RBE were proportional to LET, whereas the track average would apply if RBE were inversely proportional to LET. There are usually insufficient data in any given mutagenic system to demonstrate the precise dependence of the RBE on the LET and to decide on the most appropriate type of average. However, in a number of systems (including those for animals which are of the greatest practical relevance for this report) the data on mutagenesis reviewed in the report appear to show that RBE increases with LET, at least within the LET region up to about 100 keV/μm. Since the mean-LET values for the neutron sources used lie below 100 keV/μm, it appears that the dose-average LET would be the most useful simple parameter. This choice would have the practical convenience that the value depends little on the mean energy of the fast neutrons and most of the fast neutron data could then be classed together as high-LET data. However, 14 MeV neutrons constitute a special case, because although the mean dose-average LET is not very different from those for several other fast neutron sources (Bewley, 1968), the RBE in some biological systems is lower. The explanation may well be that these neutrons produce two main groups of ionizing secondary particles, recoil protons with a mean LET$_\infty$ around 20 keV/μm and heavier recoil nuclei as well as α particles (from the disintegration of carbon and oxygen) with a mean LET$_\infty$ around 200 keV/μm. This latter LET is well beyond the peak of biological effectiveness for most biological systems at about 100 keV/μm. Thus although the dose-average LET$_\infty$ for 14 MeV neutrons is quite high (92 keV/μm), the overall biological effectiveness is likely to be less than would be expected for this LET. There appears to be no generally applicable simple index of radiation quality for these neutrons.

Some types of neutron irradiation are not covered by the information above for fast neutrons; they are slow or thermal neutrons of intermediate energy. The slow neutrons have kinetic energies approaching thermal values and they convey energy to the irradiated medium only through nuclear capture reactions. The most frequent reactions in tissue are capture by a nitrogen-14 nucleus which leads to emission of a proton of about 0.6 MeV, a moderately-high-LET particle, capture by boron-10 giving a high-LET α particle of 2.3 or 2.8 MeV, and capture by hydrogen-1 which leads to emission of a γ ray quantum of 2.2 MeV. If the irradiated mass of tissue is small, most of the energy deposited in it is due to the protons, and the irradiation is more or less of a high-LET character. If the irradiated tissue is large, then the γ ray component of absorbed dose at any point due to the γ rays produced throughout the tissue mass outweighs the component due to protons and the irradiation is then mainly at low LET. Each case requires individual consideration. Some data have been given by Boot and Dennis (1968).

The term "intermediate energy neutrons" refers to neutrons with energies anywhere in the region between about 1 eV and a few hundred keV. The recoil particles they produce in tissue tend to have rather low energies and short ranges, often on the low-energy side of the Bragg peak of the particles. It would be difficult to predict from existing data the biological effects of such neutrons. Some experiments on their mutagenic effects have been carried out by Troitskii et al. (1965). These authors give some information about the energy distribution of the neutrons they used, but not on the LET distribution generated in the biological specimens.

For collisions of very fast neutrons or protons with the nuclei of light elements, another process occurs above an effective threshold of about 40 MeV. The fast primary particle knocks one or more fast secondary particles out of a nucleus (cascade particles) and leaves the residual nucleus with an excitation energy of 25 MeV or so. This energy is released by the emission of a few particles which may be neutrons, protons, deuterons, tritons, α particles, etc. In a photographic emulsion these relatively slow secondary charged particles produce a characteristic appearance of a few short tracks radiating out from a common centre, a so-called "star". These tracks represent a high-LET component of the energy deposition by the primary particle. The relative contribution to the total dose depends on whether the fast primary particle is charged so that it produces primary ionization (at low LET). Even if the primary particle is a neutron, some of the fast cascade particles will be protons which will produce ionization at low LET. The relative contribution of the "stars" to the total dose also depends on the thickness of matter through which the fast primary particle has already passed. The mean free path of the primary for nuclear collisions is of the order of 1 metre, and path lengths comparable to this are required for the full build-up of fast secondary (cascade) particles, which themselves can produce stars. An analysis of the dose from primary protons or neutrons up to 2 GeV has been made by Neufeld et al. (1966, 1969).

Negative π-mesons also produce stars when they interact with a nucleus, usually near the end of their range. In this case, the excitation energy of the nucleus is rather larger and about 100 MeV are available for kinetic energy of the neutrons and charged particles emitted. Details may be found in Guthrie et al. (1968).

Studies on Mutation Induction

In recent years, a wide variety of organisms has been used for investigating the induction of gene mutations and chromosomal aberrations by high-LET radiation. Results are outlined in the following sections.

MICRO-ORGANISMS

Very little relevant work has been reported in bacteria or other prokaryotes. In E. coli, Munson et al. (1970a) (see also Bridges and Munson, 1968) found that studies on high-LET mutagenesis were hampered by a progressive masking of the mutagenic effect by associated lethal effects as the LET increased. There was a steady decrease in the frequency of mutants per survivor with increasing LET, which would suggest an RBE of less than 1. It is interesting to note that Munson and Bridges (1970) and Munson et al. (1970b) have recently found a similar phenomenon in T4 phage, although it is radioresistant to inactivation. The RBE for base change mutations induced in an amber nonsense triplet decreased steadily with increasing LET. In addition, Troitskii et al. (1966) have reported an RBE of 0.6 for the induction of reversions in E. coli with 200 keV neutrons.

Troitskii et al. (1965, 1966) have reported an RBE of 45 for the induction of λ prophage in E. coli K-12 by 200 keV neutrons. They state that this was not unexpected, since RBEs reach a maximum in the intermediate-energy neutron range.

Turning to Paramecium, Kimball et al. (1959) found that post-treatment with streptomycin reduced the frequency of lethal and deleterious mutations after exposure to plutonium-239 α-particles (energy 5.16 MeV) as well as to 250 kVp x rays. The total α-particle dose of approximately 1680 rad induced about the same frequency of mutations as expected from 3 kR x rays. Thus the RBE was slightly less than 2.

FUNGI

The few experiments that have been done with high-LET radiations are on both haploid and diploid yeast, and a heterokaryon of Neurospora. The effects assayed have been killing of cells and mutation induction. Although cell-killing is known to involve the genetic apparatus in both Neurospora and yeast, only studies directly concerned with mutation induction are considered here. These have involved the induction of reversion of auxotrophic markers as well as the induction of recessive lethal mutations at specific loci (forward-mutation).

Studies to determine the RBE of thermal neutrons for mutation induction were made by Hrishi and James (1964) with a strain of diploid yeast homozygous for a histidine marker. Thermal neutrons were found to be more efficient as inducers of mutation than γ rays (Table 3). The slopes of the dose–response curves for each radiation are not linear, but increase with increasing dose. Because of this response the RBE values are dose-dependent, but range from 1.0 to 1.6 over the range of doses studied.

Additional studies on mutation induction were made on diploid yeast by Mortimer et al. (1965) who compared the genetic effects of heavy ions with those of unfiltered x rays. Reversion of a histidine and a tryptophan marker were studied. The initial portions of the dose–response curves were linear with good reproducibility from experiment to experiment; the final portions were more variable. RBEs (shown in

Table 3) were determined by comparing the slopes of the initial portions. Essentially the same results were found with both markers: with increasing LET the efficiency in the induction of reversions increases and reaches a maximum at a LET corresponding to that of boron or carbon ions of about 10 MeV/Atomic Mass Unit energy. It then decreases rapidly with a further increase in LET.

Neurospora crassa was used by de Serres et al. (1967) to study the induction of forward-mutations at specific loci. Using a heterokaryon heterozygous for two closely linked loci, *ad-3A* and *ad-3B*, they compared the genetic effects of filtered 250 kVp x rays, 40 MeV helium ions and 108 MeV carbon ions (Table 3). Higher forward-

TABLE 3
RBE Values in Experiments to Compare the Effects of High- and Low-LET Acute Irradiation on Mutation Induction in Fungal Systems

Radiations compared	Dose range (krad)	LET keV/μm	Reverse mutation	RBE Forward-mutation Overall	Point mutation	Chromosome deletion	Reference and organism
^{60}Co γ rays	0.25–40						Hrishi and James (1964)
Thermal neutrons		200	1.0–1.6	—	—	—	(*Saccharomyces cerevisiae*)
50 kVp x rays	5–20						
10.2 MeV deuterium ions		2.5	1.0–1.5	—	—	—	Mortimer et al. (1965);
9.8 MeV helium ions		9.8–10.0	1.3–1.4	—	—	—	Nakai and
9.7 MeV lithium ions		22–24	1.5–2.0	—	—	—	Mortimer (1967)
8.9 MeV boron ions		68–75	1.9–2.5	—	—	—	(*Saccharomyces*
8.4 MeV carbon ions		100–120	1.7–2.6	—	—	—	cerevisiae)
7.1 MeV neon ions		300–340	0.7–1.0	—	—	—	
250 kVp x rays	1–40						
40 MeV helium ions		18.2–19.0	—	1.8	—	—	de Serres et al. (1967)
108 MeV carbon ions		179–210	—	5.0	—	—	(*Neurospora crassa*)
250 kVp x rays	1–40						
40 MeV helium ions		18.2–19.0	—	—	2.2	5.5	de Serres (1970)
108 MeV carbon ions		179–210	—	—	5.5	73.8	(*Neurospora crassa*)

Note: The RBEs for point forward-mutations found by de Serres *et al.* are not significantly different from corresponding RBEs for overall forward-mutations, but those for chromosome deletions are.

mutation frequencies were obtained with comparable doses of both heavy ions than with x rays. The RBEs for overall induction of *ad-3* mutations are 1.8 for helium ions and 5.0 for carbon ions.

Genetic analysis of the *ad-3* mutations recovered in the previous experiments by de Serres and co-workers (1967) makes it possible to characterize the specific locus mutations and to distinguish point mutations (which alter genes but do not remove them) from chromosome deletions (in which genetic material is lost). As a result, RBE values can be determined for each class of genetic alteration (Table 3 and de Serres, 1970). These data show that the efficiency of the higher-LET radiation relative to low LET is greater for the induction of recessive lethal mutations by chromosome deletion than by point mutation. However, point mutations were induced much more frequently than deletions and therefore very largely determined the overall RBE. The difference in the RBEs for helium and carbon ions reported in this investigation from those reported for the comparable experiments with diploid yeast may result from the fact that in the yeast experiments only point mutation could be studied and not chromosome deletion. The high RBE in *Neurospora* for induction of

recessive lethal mutations by chromosome deletion with carbon ions would be completely missed with most assay systems.

The nature of the genetic alterations resulting in point mutations at the *ad-3B* locus by x rays has been studied in experiments on revertibility after treatment with specific chemical mutagens (Malling and de Serres, 1967a and unpublished). A comparison of the spectra of allelic complementation among x ray, helium-ion and carbon-ion induced *ad-3B* mutants shows no significant difference. Since a correlation has been found between complementation pattern and genetic alteration (Malling and de Serres, 1967b), we can conclude that x rays, helium ions and carbon ions produce the same spectrum of genetic alterations at the molecular level resulting in point mutations.

HIGHER PLANTS

This section deals with mutagenic effects (both somatic and hereditary) of high-LET radiations on higher plants as reported in papers published during the period 1961–70. Results prior to this were discussed in the 1963 report of the RBE Committee to ICRP and ICRU and have been reviewed extensively by Bora (1961), by Gopal-

TABLE 4
RBE VALUES FOR MUTAGENESIS OF FAST ("FISSION") NEUTRONS VS. LOW-LET RADIATIONS ON PLANT
SPECIES IRRADIATED AS DRY SEEDS

Material and species irradiated	Criteria (response level, generation)	Low-LET radiation	RBE	Reference
Seeds—wheat *Triticum*, $2\times, 4\times, 6\times$ spp.	Chromosome aberrations (R_1)	x	5–9	Bhatt *et al.* (1961)
	Chlorophyll mutations (R_2)	γ	25	Matsumura (1966)
Seeds—maize *Zea mays*	Mutation (1/leaf, R_1)	x	76	Smith (1969)
	Mutation (2/leaf, R_1)	x	37	Smith *et al.* (1968)
	Albino mutations (1%, R_2)	γ	2	Stoilov (1968)
	Chlorophyll mutations (1%, R_2)	γ	3–7	Stoilov (1968)
	Total mutations (1%, R_2)	γ	9–15	Stoilov (1968)
Seeds—*Arabidopsis thaliana*	Single locus mutation (1%, R_1)	γ	16	Fujii (1964a, 1969)
Seeds—*Nigella damascena*	Chromosome aberrations (R_1)	γ	78	Moutschen *et al.* (1969)

TABLE 5
RBE VALUES FOR MUTAGENESIS OF MONOENERGETIC AND MIXED SPECTRUM NEUTRONS VS. LOW-LET
RADIATIONS IN HIGHER PLANTS

Material and species irradiated	Criteria (generation scored)	High-LET Energy	High-LET LET keV/μm	Low LET	RBE	Reference
Seeds—maize	Single locus mutation (R_1)	0.43 MeV	72	x	100	Smith (1967);
Zea mays		0.65 MeV	67	x	88	Smith *et al.* (1964,
		1.00 MeV	58	x	81	1966)
		1.50 MeV	47.5	x	69	
		1.80 MeV	42.5	x	69	
Seeds—*Nigella damascena*	Chromosome aberrations (R_1)	0.43 MeV	72	γ	115	
		0.65 MeV	67	γ	67	Moutschen *et al.*
		1.00 MeV	58	γ	85	(1969, 1970)
		1.50 MeV	47.5	γ	61	
		1.80 MeV	42.5	γ	59	
Inflorescence— *Tradescantia* species (Stamen hairs)	Single locus mutation (R_1)	0.65 MeV	67	x	18–40	Davies and Bateman (1963)
	Pink cells (R_1)	0.43 MeV	72	x	13–31	Underbrink *et al.* (1970, 1971)
Seeds—rice *Oryza sativa*	Chlorophyll mutations (R_2)	4.7 MeV	31	γ	19	Matsumura and Mabuchi (1965)
Seeds—*Nigella damascena*	Chromosome aberrations (R_1)	Thermal	~150	γ	81	Moutschen and Moutschen-Dahmen (1970)
Inflorescence—*Tradescantia paludosa*	Chromosome aberrations (R_1)	3 MeV (+ fission)	40	γ	100 (max)	Neary *et al.* (1963)
Inflorescence— *Tradescantia* spp.	Pink cells (R_1)	80 keV*	Est. 80*	x	10–16	Underbrink *et al.* (1971)
Pollen—tomato *Lycopersicon esculentum*	Specific locus mutations (R_2)	Thermal $^{14}N(n,p)^{14}C$	45	γ	1.5–3.0	Ecochard (1970)
Root tip cells *Allium* var. caba	Chromosome breaks (R_1)	200 KeV	85	γ	100	Troitskii *et al.* (1966)
Seeds—wheat *Triticum aestivum*	3 specific loci (R_1)	2 MeV	49	γ	20	Rana and Swaminathan (1967)
	Chlorophyll and viable mutations (R_2)	2 MeV	49	γ	20	
Seeds—wheat *Triticum monococcum*	Single locus mutation (R_1)	14 MeV	(16)(222)	γ	13	Fujii (1964a, 1969)
	Chlorophyll mutations (R_2)	14 MeV	(16)(222)	γ	15	Matsumura (1966)
	Chlorophyll mutations (R_2)	4.7 MeV	31	γ	40	Matsumura (1966)
Seeds—maize *Zea mays*	Single locus mutation (R_1)	14 MeV	(16)(222)	x	49	Smith and Rossi (1966)
Seeds—Arabidopsis	Single locus mutation (R_1)	14 MeV	(16)(222)	γ	15	Fujii (1964b)

* Spectrum of energies from near zero to about 120 keV, but mostly from 100 down to 1 keV.

TABLE 5—*Continued*

| Material and species irradiated | Criteria (generation scored) | Irradiations High-LET | | Low LET | RBE | Reference |
		Energy	LET keV/μm			
Pollen—maize *Zea mays*	Single locus mutation (R_1)	14 MeV	(16)(222)	γ	5	Fujii (1964b)
Root cells—broad bean *Vicia faba*	Chromatid aberrations (R_1)	14 MeV	(16)(222)	x	3–5	Savage (1968)

TABLE 6
RBE VALUES FOR MUTAGENESIS OF HIGH-LET RADIATIONS PRODUCED BY ACCELERATED HEAVY IONS

| Material and species irradiated | Criteria (generation scored) | Irradiations High-LET | | | Low LET | RBE | Reference |
		Particle	Incident energy	LET keV/μm			
Seeds—*Arabidopsis thaliana*	Single locus mutation (1%, R_1)	^{4}He	39.8 MeV	20	γ	10	Fujii (1969);
	Single locus mutation (0.5%, R_1)	^{12}C	108 MeV	230	γ	35	Fujii *et al.* (1966, 1967)
	Single locus mutation (0.5%, R_1)	^{40}Ar	277 MeV	2 500	γ	5	
	Single locus mutation (3%, R_1)	^{4}He	39.8 MeV	18	x	1.4	
		^{4}He	9.1 MeV	74	x	21	
		^{7}Li	17.4 MeV	172	x	15	
		^{12}C	44.0 MeV	409	x	12	Hirono *et al.* (1970)
		^{16}O	67.0 MeV	752	x	11	
		^{20}Ne	91.2 MeV	1 030	x	1.3	
		^{40}Ar	222 MeV	1 890	x	1.6	
Seeds—wheat *Triticum monococcum*	Chromosome aberrations (R_1)	$\alpha+^7$Li	(1.6)(0.9) MeV	~170	γ	23	Matsumura *et al.* (1963)
	Chlorophyll mutations (R_2)	$\alpha+^7$Li	(1.6)(0.9) MeV	~170	γ	29	
Seeds—maize *Zea mays*	Single locus mutation	Mesons	8 GeV	Secondary	x	3	Micke *et al.* (1964a, b); Smith (1967)
		Protons	28 GeV	Secondary	x	3–5	Smith (1967); Smith *et al.* (1965)
Seeds—*Nigella damascena*	Chromosome aberrations (R_1)	Protons	2.8 GeV	Secondary	γ	2	Moutschen *et al.* (1969)

Ayengar and Swaminathan (1964) and by Smith (1962). This earlier work showed that RBE values for chromosome aberrations depended both on the dose given and the type of aberration studied. Very high values could be obtained when the dose was protracted because of the decreased yield from low-LET irradiation under these conditions. Fission neutrons were found to be more effective than 14 MeV neutrons.

Relevant information about the more recent experiments and computed RBE values for fission neutrons are summarized in Table 4, for monoenergetic and mixed spectrum neutron energies in Table 5, and for high-LET radiations produced by accelerated heavy ions in Table 6. Dose average LETs are listed in Tables 5 and 6. A wide variety of plants has been used as experimental materials for mutation studies. Some of the more favoured species have been *Zea mays* (maize), *Triticum* spp. (wheat), *Tradescantia* spp. (spiderwort), *Oryza sativa* (rice), *Arabidopsis thaliana* (a short life cycle crucifer), and *Nigella damascena* (devil-in-the-bush).

The plants have been irradiated at different stages of development, thus involving different tissues: seed, inflorescences and pollen. The stage or plant part irradiated seems not to influence consistently the magnitude of the RBE. Seeds have been preferred experimental materials since they can be subjected, without impairment of function, to a wide range of radiation doses and of environmental conditions (temperature, moisture, atmospheric content and storage time). These factors greatly influence the degree of response, especially with low-LET radiations and particularly for seed irradiations where large differences in moisture or oxygen content may exist. For example, Smith and Combatti (1967) found for a single locus somatic mutation in maize that seeds of 6.7% moisture gave an RBE for fission neutrons vs. 250 kVp x rays of 67, whereas after 36 hr of soaking prior to irradiation the RBE was reduced to 5. These changes in RBE were due largely to changes in sensitivity to low-LET radiations. Thus a listing of RBE values is of little meaning unless the state of the plant materials and the environmental condi-

tions under which the irradiations were carried out are controlled and specified. Although this information is not completely available for the experiments listed in the tables, nevertheless the irradiations were usually carried out in air, and irradiated seeds were usually described as dry and dormant, which means around 10 to 20% moisture. It can be safely assumed that seeds were not stored but were in all probability sown soon after irradiation in moist soil under oxic conditions.

All irradiations listed in Tables 4, 5 and 6 were administered as acute single exposures apart from those of Ecochard (1970) and Neary *et al.* (1963) listed in Table 5. The low-LET radiations used for comparison with high-LET were either 250 kVp x rays or γ rays, the latter most frequently from ^{60}Co but in some cases ^{137}Cs. In Table 4 the fission neutrons were unmodified or slightly degraded, with a spectrum of energies peaking at 1 to 1.5 MeV. Most of the work on thermal neutrons was not included in the tables because the absorbed dose based on the elemental composition of the plant material was not computed. In Table 6 the high-LET radiations are mainly accelerated heavy ions (^{4}He to ^{40}Ar); but also high energy negative π mesons and protons which gave RBEs above unity probably due to high-LET secondary particles from nuclear disintegrations.

The types of response used to measure the relative mutagenic effectiveness of high- vs. low-LET radiation have been scored in either the treated generation (R_1) or its progeny (R_2 generation). Those scored in the treated generation include chromosome and chromatid aberrations, as well as somatic mutations arising mainly from loss of a dominant marker in tissues of a heterozygous plant. Those scored in the progeny include direct measures of germinal mutation at loci governing chlorophyll production.

Effects of dose and intensity

Since response to high-LET radiations has for the most part an exponential or straight line relation to dose, whereas low-LET radiations show a power function or curvilinear relation, the magnitude of the RBE will vary with dose. In turn, the dose required to produce an effect will depend on the radiosensitivity of the genotype, the criterion of response scored, and the level of damage reached. Most mutations are scored at a level of 1 to 3%, based on numbers of mutant plants in the R_1 generation or those giving mutants in the R_2. Consequently, the criteria in the tables represent comparable levels of response to a considerable extent.

The wide range in tolerance to radiation shown by seeds of higher plants, and the widely different radiosensitivities of higher plant species, have made it possible to explore the relative biological effectiveness of high- vs. low-LET radiations over an extremely wide range of doses. The relationship between the doses of high- and low-LET radiation producing equal effects can be expressed by a simple power function, perhaps more significantly than by a series of changing RBE values. For instance, when dry seeds of maize were irradiated, it was found that log $N = 2.6 + 1.95$ log X, where N and X are the fission neutron and 250 kVp x-ray doses required to produce equal effects with respect to nine different criteria, two being specific somatic mutations (Smith, 1969; Smith et al., 1968). This relation indicates that for all these criteria the effect is determined by the first power of the neutron dose and by approximately the square of the x ray dose.

In general, higher RBEs than those quoted in the tables would be expected at low doses and dose rates, because of the usual dose-squared component in the low-LET response. The magnitude of the increase will depend greatly on the mutational end-point used. Underbrink et al. (1970) found that the RBE for induction of a genetic change (probable deletion) from blue to pink cells in Tradescantia stamen hairs rose from 12.8 at a 0.43 MeV neutron dose of 9.2 rad to 30.7 at a neutron dose of 0.45 rad. The increase was even more pronounced with certain other "aberrant events" (e.g. formation of Blue Giant cells) in which the nature of the change was less clear. Neary et al. (1961, 1963) found

very high RBEs for various types of chromosome aberrations in Tradescantia when the exposure times were protracted. They deduced that RBEs would reach maximum values of the order of 100 with very long exposures, because the dose-squared (interaction) component would tend to vanish while the linear one remained small.

Smith and Rossi (1966) found that the response of dry seeds to low-LET radiation was independent of dose rate over a range of 1 758 to 10.3 rad/min. However, a still lower exposure rate of 1 500 R per 20 hr day (1.25 R/min) was more effective with dry seeds than equivalent acute doses and this was attributed to cumulative effects of long-lived free radicals (Natarajan and Maric, 1961). In a "wet" system, i.e. plantlets of African violet (Saintpaulia) propagated from leaf cuttings (Broertjes, 1968), mutation frequency per absorbed dose was affected by dose rate of both fast neutrons and x rays, and for the latter reached a peak at about 200 rad/min.

A number of mutation experiments with higher plants have covered a range of LETs and thus provide useful evidence on the effect of LET on the RBE for mutagenesis. Thus the irradiation of seeds by monoenergetic neutrons has shown that RBEs from the induction of both chromosomal aberrations and single locus mutations in somatic cells increase markedly with increase of LET from 42.5 keV/μm to 72 keV/μm, as shown in Table 5. The use of accelerated heavy ions has allowed a very wide range of LETs to be explored by Hirono et al. (1970) who obtained a peak RBE for induction of single locus somatic mutations at a LET of 74 keV/μm (Table 6). Figure 2 shows that the authors obtained remarkably similar RBE–LET relationships with respect to somatic mutation, tumour induction and growth inhibition. Unfortunately, there have been no similar experiments involving mutations transmitted to the next generation, but Matsumura (1966) has shown that fission neutrons are more effective than 14 MeV neutrons for the induction of chlorophyll mutations, scored in the next generation (see Tables 4 and 5). RBE values for transmitted mutations

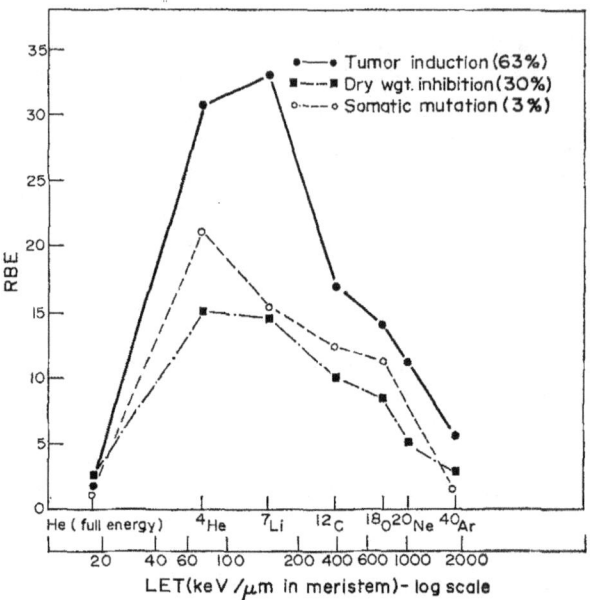

Fig. 2. RBE values for accelerated heavy ions and 250 kVp x rays at different LETs (shown on the abscissa) with respect to tumour induction, growth inhibition and the induction of somatic mutations in seeds of *Arabidopsis*. With each of the ions, except He (full energy), the Bragg peak was adjusted to fall in the shoot meristem of the seed. From Hirono *et al.* (1970); semi-log plot.

generally seem lower than for those scored in the irradiated generation, which tend to be much higher than for other organisms studied. It is clear that in higher plants the RBE values found depend greatly on the test system used and the conditions of the experiment, as well as on the LET of the radiation.

Almost the only evidence from higher plants for differential specificity of mutagenesis by high- vs. low-LET radiations, comes from barley. The *eceriferum* mutants in this species are characterized by absence or reduction of the wax coating on the epidermis (Lundqvist, 1967; Lundqvist *et al.*, 1968). The mutants have been tested for allelism by diallelic crosses and two mutants are considered allelic when the F_1 has an *eceriferum* phenotype. The 93 observed mutants induced by sparsely ionizing radiations (x and γ rays) were distributed rather evenly among 32 differently located *eceriferum* loci. In contrast, the mutants produced by densely

ionizing radiations (protons, neutrons and α particles) showed a prominently skewed locus distribution in that about 30% of the 53 mutants produced at 18 loci were found to be located at a particular single locus. There is no evidence to explain the preferential effectiveness of high ionization density for this locus. A rather similar phenomenon was found previously in the *erectoid* mutations of barley by Hagberg *et al.* (1958) and Persson and Hagberg (1969).

An example of a differential effect of LET on chromosome aberrations is provided by Savage *et al.* (1968) who showed that α particle irradiation of *Tradescantia* pollen produced an increased incompleteness of isochromatid intrachanges compared with radiations of lower LET.

INSECTS

In recent years information on the relative genetic effectiveness of high- and low-LET radia-

TABLE 7
RBE OF FISSION NEUTRONS (1.5 MeV) AND γ RAYS AT DIFFERENT STAGES
OF GERM-CELL DEVELOPMENT IN THE SILKWORM FOR THE INDUCTION
OF SPECIFIC-LOCUS MUTATIONS. COMPARISONS MADE AT THE 10^{-3}/LOCUS
MUTATION FREQUENCY LEVEL. FROM MURAKAMI *et al.* (1965)

Type of cell	RBE		
	pe	*re*	mean
Primordial spermatogonia in newly hatched larvae	1.7	1.9	1.8
Late spermatogonia in 7-day larvae	4.2	3.5	3.9
Primordial oogonia in newly hatched larvae	2.1	2.4	2.3
Late oogonia in 7-day larvae	3.8	3.0	3.4

tions in insects has come mainly from work on the silkworm, although there have been additional data from *Drosophila* and from the hymenopteran *Dahlbominus*.

Most of the silkworm studies, like those in the mouse, have been concerned with the induction of specific locus mutations. The two loci concerned affect egg colour, the tester stock used being homozygous for the two linked recessive genes pink egg (*pe*) and red egg (*re*).

Murakami *et al.* (1965) have studied the mutagenic effectiveness in male and female germ cells of fission neutrons of mean energy 1.5 MeV (which show an exponential response) and have found similar mutagenicities in the two sexes (Table 7). For RBE calculations, comparisons with γ radiation were made at the 10^{-3}/locus level of mutation frequency. On average, fission neutrons were 1.7 times as effective as 14 MeV neutrons for these four comparable stages. The reason for the exponential rather than linear type of neutron dose–response curve in these and other silkworm experiments is not known, but may well be connected with germinal selection.

Machida and Nakao (1969) have compared specific-locus mutation frequencies in pupal prophase I oocytes after irradiation with 2.5 MeV neutrons and with 200 kVp x rays. The dose–response curve for x rays was exponential, but at lower doses the RBE was about 3. Unfortunately

there are no comparable data for other neutron energies.

Table 8 summarizes the results obtained on male and female germ cells, mainly immature, with 14 MeV neutrons at dose rates between 1 and 40 rad/min compared with those for ^{137}Cs γ irradiation at 100 rad/min. The γ dose–response curves for the various germ-cell stages were approximately linear, but those for neutrons were not, except when spermatozoa were irradiated. The ratios of doses required to give a mutation frequency of 10^{-3}/locus are compared in Table 8 also; if a lower frequency had been chosen the RBE values would have been lower. It can be seen that estimated values are all lower than those for irradiation of comparable stages with fission neutrons.

Table 8 shows that RBEs depend on developmental stage, but this was largely due to variations in the gamma rather than the neutron response. It is suggested by Murakami *et al.* (1964, 1965) that there may be differing capacities for repair of γ-induced mutational damage in different stages but probably no repair of neutron damage (see also Kondo, 1965). It can be seen that RBEs are higher in later than in earlier stages, with the exception of primordial spermatogonia in eggs, and reach a maximum with irradiation of sperm. At low doses oogonia and spermatogonia gave similar mutation frequencies.

TABLE 8

RBE OF 14 MeV NEUTRONS AND γ RAYS AT DIFFERENT STAGES OF GERM-CELL DEVELOPMENT IN THE SILKWORM
FOR THE INDUCTION OF MUTATIONS AT THE *pe* AND *re* LOCI. COMPARISONS MADE AT THE 10^{-3}/LOCUS LEVEL OF
MUTATION FREQUENCY, EXCEPT FOR MATURE SPERM (WHOLE DOSE RANGE)

Type of cell	RBE			Reference
	pe	*re*	mean	
Primordial spermatogonia in hibernating eggs	1.8	2.9	2.4	Murakami and Tazima (1965)
Primordial spermatogonia in newly hatched larvae	0.8	1.0	0.9	Murakami and Kondo (1964)
Late spermatogonia in 7-day larvae	3.2	2.1	2.7	Ibid.
Mature sperm in late pupae	5.3	6.7	6.0	Murakami (1966, 1970)
Primordial oogonia in newly hatched larvae	1.2	1.2	1.2	Murakami and Kondo (1964)
Late oogonia in 7-day larvae	1.7	2.8	2.3	Ibid.

Evidence that neutron-induced mutational lesions are but little susceptible to repair has been obtained from experiments involving post-irradiation treatment of silkworm spermatogonia with the base analogue 5-bromodeoxyuridine (BUDR). With x-irradiation BUDR enhanced the yield of specific locus mutations about 2–3 times at the maximum (Murakami and Tazima, 1963), but there was very little effect with 14 MeV neutrons (Murakami, 1967). Murakami and Ito (1969) considered that the small size of the enhancement effect with neutrons was because more double-strand breaks, not susceptible to repair, would be induced by densely ionizing radiations. In line with this view are the findings of Tazima *et al.* (1968) that a much higher proportion of mosaic mutations (not affecting the whole body) are recovered after treatment with chemical mutagens than with 14 MeV neutrons. The predominantly whole-body mutations obtained with the latter treatment are believed to result from lesions affecting both strands of the DNA double helix.

Recently Murakami (1970) reported that five different silkworm strains, in which LD_{50}s for embryonic mortality ranged from 140 R to 1 710 R, also showed considerable differences in their specific-locus mutation rates and in result-ant RBEs, after x and fission-neutron irradiation of gonia. The strains with the greatest sensitivity to radiation-induced killing were also those giving the highest mutation frequencies and the lowest values for the RBE of fission neutrons relative to x rays.

Murakami *et al.* (1966) have shown that fractionation of a dose of 14 MeV neutrons can more than double the frequency of specific locus mutations in spermatogonia irradiated soon after hatching of the silkworm egg. This is similar to previous findings by Tazima and Murakami (1963) after low-LET irradiation, but the peak yield was obtained with a 36-hr interval for neutrons, in contrast to 18 hr for x and γ rays. Oogonia were affected less. The enhancing effect is probably the result of differential radiosensitivity within the gonial cell cycle.

The induction of dominant lethals in silkworm germ cells by 14 MeV and fission neutrons has also been studied by Murakami (1968, 1970). With 14 MeV neutrons and ^{137}Cs γ rays, RBEs of 1.6, 4.4 and 8.2 were found for primordial germ cells, spermatogonia in larvae and mature spermatozoa respectively. A linear dose–response relationship existed only with spermatozoal irradiation; other germ cells have been compared at the 50% survival level. At the same level the

RBE of fission neutrons relative to γ rays was 11.2 with spermatogonial irradiation, 2.5 times the value for 14 MeV neutrons. Thus the general pattern is very similar to that for induction of specific locus mutations, namely higher RBEs with fission neutrons than with 14 MeV neutrons and with more mature than with less mature cells.

One great advantage of the hymenopteran *Dahlbominus* for mutational studies is that un-mated females produce only haploid male progeny in which all mutations are expressed. Those particularly studied by Baldwin (1968) and co-workers have been four classes of eye-colour mutants (carmine, claret, chestnut and russet), which arise at a minimum of eight loci. Baldwin and Cross (1966) compared the frequencies of such eye-colour mutations in female *Dahlbominus* of different ages after irradiation with 750 rad of 14.6 MeV neutrons (80 rad/min) or of ^{60}Co γ rays (100 rad/min). Mutation frequencies rose with age of the insects at the time of exposure, i.e. with increasing proportions of mature oocytes. Neutron/gamma RBEs (ratios of mutation frequencies) were 1.2–1.4. In separate experiments Baldwin (1968) showed that when oocytes were irradiated at a stage of constant radiosensitivity the yield of eye-colour mutations was only slightly higher from chronic than from acute γ irradiation. There is some evidence that germinal selection was not responsible for the lower yield with acute irradiation. The findings suggest that with chronic irradiation of oocytes in *Dahlbominus* the RBE for 14 MeV neutrons vs. γ rays may not differ greatly from unity.

Rather similar work to Baldwin's was carried out by Kayhart (1956) on *Mormoniella vitripennis*, another hymenopteran showing haplo-diploidy. This paper was not discussed in the RBE report but deserves attention. Like Baldwin, Kayhart irradiated virgin females and looked for eye-colour mutations in their haploid sons; however, effects of thermal neutrons, fission neutrons from detonation of nuclear devices and acute x irradiation were studied rather than 14 MeV neutrons and γ rays.

Kayhart reported that the RBE for fission neutrons vs. x rays was 17–21 at low doses (below 100 rad neutrons) falling to 2–4 at higher ones. No RBEs were calculated for thermal neutrons. Kayhart considered that the decreased relative effectiveness of fast neutrons at high doses was to be expected if many of the mutations were due to minute rearrangements and deletions.

Some earlier studies on specific-locus mutation frequencies in *Drosophila* need to be quoted here because of their comparability with work in other insects and in mice. Mickey (1954) compared mutation frequencies at 8 specific loci in *Drosophila* spermatozoa given 3 000 R 250 kVp x irradiation, 1 000 rep* cyclotron neutrons (peak energy about 1 MeV) and various doses of nuclear-detonation neutrons (also around 1 MeV). On the assumption of linear dose–response relationships the average RBE for neutrons vs. x rays was estimated to be 4.0 for the cyclotron neutrons and 4.5 for the nuclear-test neutrons. The neutrons had an RBE of 2 relative to x rays for the induction of sex-linked recessive lethal mutations. Ives *et al.* (1954) found that nuclear-detonation neutrons from an atomic explosion were about four times as effective as 120 kVp x rays for the induction of specific locus and other visible mutations in *Drosophila* spermatozoa.

Oster (1963) has studied mutations induced in mature spermatozoa (inseminated females) at the dumpy (*dp*) locus. Doses of 4 000 rad acute x rays, 4 000 rad acute ^{60}Co γ rays and 750 rad fission neutrons induced *dp* locus mutations at frequencies of 0.2%, 0.5% and 0.2% respectively. It is clear that the RBE for the fission neutrons relative to low-LET radiation was above unity, but no accurate figure can be given on the data available. Unexpectedly, the frequency of mosaic mutations was around 30% in all three series, which suggested that both DNA strands were affected to the same extent in radiations of very different LET.

Nakao and Machida (1968, 1970) have also studied mutation induction in *Drosophila* sperm-

* Probably about 930 rad.

TABLE 9

NEUTRON VS. X-RAY RBEs FOR THE INDUCTION OF RECESSIVE LETHALS BY ACUTE IRRADIATION OF *Drosophila* GERM CELLS: SOME RECENT RESULTS

Radiation energies		Doses in rad		Type of lethal scored	Treated germ cells	RBE	Reference
Neutron (MeV)	X ray (kVp)	Neutron	X ray				
0.2–0.3	250	267–1 066	9 60–3 840	Sex-linked	Oogonia Oocytes	2	Dickerman (1967)
0.7	250	200–1 000		2nd chromosome	Post-meiotic male Pre-meiotic male	2.2 2.1	Lamb *et al.* (1967)
2	150	245–1 460	960–2 800	Sex-linked	Spermatozoa Spermatids	1.2–2.2	Dauch *et al.* (1966)
2.5	200	500–3 700	930–2 790	Sex-linked	Spermatozoa Spermatids	2.3–3.7	Nakao and Machida (1968, 1970)
15	250	1 200–4 000	1 600–4 000	Sex-linked (ring X)	Spermatozoa Late spermatids	0.8 1.2	Sobels (1967) Sobels and Broerse (1970)

atozoa at the *dp* locus, but with 2.5 MeV neutrons. They obtained a value of 2.1 for the RBE of neutrons vs. x rays.

There have been several recent studies on neutron RBEs for induction of autosomal and sex-linked recessive lethals in *Drosophila*; these are summarized in Table 9. Again, there is evidence for an increased effectiveness of fission neutrons over those of higher energies; in addition, RBEs tended to be higher with spermatid than with spermatozoal irradiation. Only in the experiments of Lamb *et al.* (1967) were spermatogonia sampled; the RBE of 2.05 can be compared with the mean value of 3.9 obtained by Murakami *et al.* (1965) for the specific-locus mutation frequency after fission neutron irradiation of late spermatogonia in the silkworm (see Table 7).

Little further information on neutron RBEs for induction of chromosomal aberrations has accrued since the RBE Report. As Table 10 shows, a fairly wide range of values has been found, the lowest ones again being associated with 15 MeV neutrons. A wide variety of mating techniques was used and it is not always possible to separate clearly the results of spermatozoal and of spermatid irradiation, so a range of values is given. In general, however, spermatid irradiation gave the higher RBE values, mainly because of lower mutation frequencies with low-LET radiation. There is a useful discussion on this subject by Sobels and Broerse (1970).

Traut (1963) compared his results for translocation frequency after x irradiation of *Drosophila* spermatozoa (treated in females) at low doses with those obtained by Muller (1954) for fission neutron irradiation of males and concluded that the RBE at low doses was 4.5 to 5.9 depending on dosimetric criteria. He thought that correction for the slightly different germ-cell stage studied might increase the value somewhat. Nakao and Machida (1968, 1970) found that the RBE of 2.5 MeV neutrons for dominant lethal induction in spermatozoa increased markedly with decreasing dose, so that a higher value than that given in Table 10 would be expected at low doses. Sobels and Broerse (1970) found that for neutrons vs. x rays the RBEs for induction of translocations in late spermatids increased at lower doses, because of the linearity of the neutron response, while the x ray yield increased with a higher power of the dose than unity.

TABLE 10

NEUTRON VS. X-RAY RBEs FOR INDUCTION OF CHROMOSOME ABERRATIONS IN *Drosophila* AFTER ACUTE IRRADIATION OF SPERMATOZOA AND SPERMATIDS

Radiation energies		Doses in rad		Aberration studied	RBE	Reference
Neutron	X ray	Neutron	X ray			
2 MeV*	150 kVp	245–1 460	1 000–4 000	Translocations (II, III)	1.7–3.3	Dauch *et al.* (1966)
2.5 MeV	200 kVp	500–3 700	930, 2 790	Translocations (II, III, Y)	5.6–5.7	Nakao and Machida (1968, 1970)
2.5 MeV	200 kVp	500–3 700	930, 2 790	Dominant lethals	2.8–3.6	Nakao and Machida (1968)
2.5 MeV	200 kVp	100–2 500	500–5 000	Dominant lethals	3.5 (50% survival)	Nakao and Machida (1970)
15 MeV	250 kVp	1 200–4 000	1 600–4 000	Translocations (II, III)	1.0 (spermatozoa) 1.1–2.3 (late spermatids)	Sobels (1967) Sobels and Broerse (1970)

* Mean energy actually about 4 MeV (see Sobels and Broerse, 1970).

In spermatozoa, however, x rays gave a linear response.

It will be noticed that there is still a dearth of information on RBEs for the induction of chromosomal aberrations in the important premeiotic stages, no doubt because of the very low aberration frequencies associated with these stages. However, Panikovskaya and Troitskii (1968) found that intermediate (200 keV) neutrons showed about the same effectiveness as γ rays for the induction of x chromosomal deletions when spermatozoa or spermatogonia were irradiated, but were more effective at the most radiosensitive stages (spermatids and spermatocytes). Their effect on fertility was less than that for γ rays or fission neutrons.

In conclusion, recent work on insects suggests that for recessive lethal and visible mutations, RBEs for fission neutrons are higher than the figures suggested in the RBE Report (1963), namely "between 1 and 2" and "not far removed from unity" respectively. However, the general picture remains similar in showing higher RBEs for fission than for 14 MeV neutrons. For chromosomal aberrations, the shape of the dose–response curves indicates that RBEs will tend to increase with decreasing dose, except with spermatozoa, while they also tend to increase with decreasing neutron energy, from 15 MeV to the fission energy spectrum.

EXPERIMENTAL MAMMALS

All recent work in this field has used the mouse as experimental mammal; most of it has been concerned with the germ cells most at risk, namely spermatogonia in the male, and oocytes arrested at the dictyate stage in the female. The induction of specific-locus recessive mutations, dominant visible mutations and chromosomal changes has been studied in some detail, so that the overall picture is much clearer than at the time of the RBE Report (1963).

As Table 11 and Figs. 3 and 4 show, high-LET radiation in the form of fast fission neutrons has proved much more effective than low-LET for inducing specific-locus mutations in mouse spermatogonia over a wide range of dose rates. The neutron dose rate has little, if any, effect on mutation frequency, but the dose rate of the low-LET radiation has a marked effect (Russell *et al.*, 1958), since acute x irradiation of spermatogonia induces, on the average, 3.3 times as many mutations per rad than chronic γ irradiation (Russell, 1965a). If the RBE value of 5.8 for neutrons vs. acute x rays (Table 11) is multiplied

TABLE 11
RBEs OF FISSION NEUTRONS AND LOW-LET RADIATIONS FOR THE INDUCTION OF SPECIFIC LOCUS MUTATIONS
IN MOUSE SPERMATOGONIA

Dose in rad		Neutron dose rate (rad/min)	Mutation frequency $\times 10^5$	Low-LET comparison	RBE	Reference
Neutron	Gamma					
52	7	79	8.5 ⎫			
52	7	0.8	10.1 ⎬	Acute x	5.8	Russell (1965a)
55	8	0.2	10.2 ⎭			
88	13	0.1	14.6	Chronic γ	18	Russell (1965a)
214	93	0.002	22.9 ⎫	Chronic γ	23 ⎫	Batchelor *et al.* (1966, 1967)
62	42	0.001	9.9 ⎭		17 ⎭	Searle (1967)

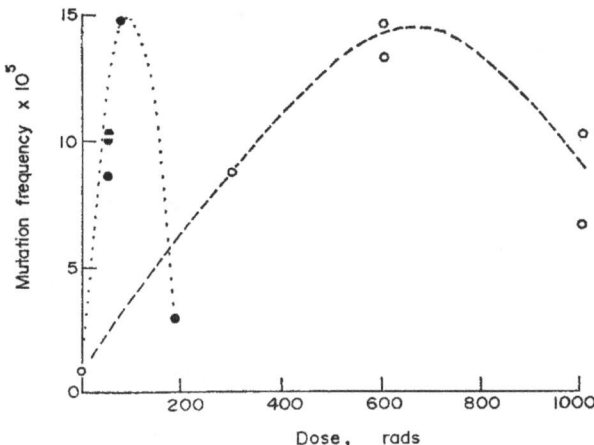

FIG. 3. Specific-locus mutation frequencies in mouse spermatogonia after fast neutron exposures of up to 12 hr (solid circles) and acute x ray exposures (open circles). The γ components of the neutron doses have been omitted. The lines have been drawn by eye and are for guidance only. Note the falling off in yield at higher doses. Combined data from Batchelor *et al.* (1967), Phillips (1961), Russell (1963) and Russell (1965a).

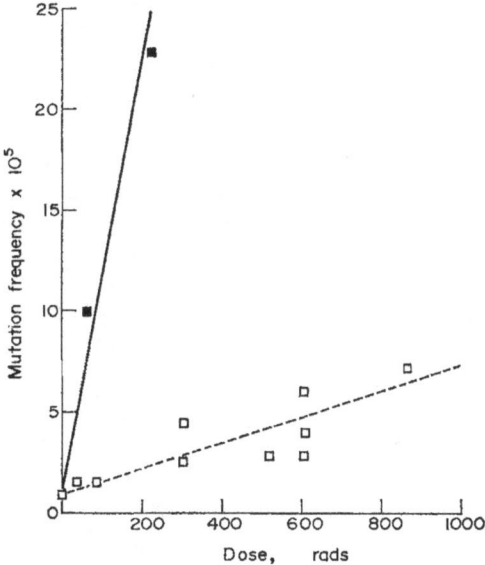

FIG. 4. Specific-locus mutation frequencies in mouse spermatogonia after chronic exposures to fast neutrons (solid squares) and to γ rays (open squares), with straight lines of best fit (Searle, 1967). The γ components of the neutron doses have been omitted. Combined data from Batchelor *et al.* (1966, 1967), Carter *et al.* (1958), Phillips (1961) and Russell (1963).

by this factor of 3.3 we obtain a figure very close to the direct estimates of RBE for neutrons vs. chronic γ rays; thus the various sets of data are in good agreement.

Some further information on the effects of dose rate on the frequency of induction of specific-locus mutations in mice by low-LET radiation may be useful. With spermatogonial irradiation, Russell (1963) found evidence for an intermediate mutational response when the x ray exposure rate was decreased from 90 R/min to 9 R/min (total exposure of 600 R), while the full reduction in yield was observed at γ-ray exposure rates of 0.8 R/min and below. With irradiation of maturing oocytes, however, the yield after 0.8 R/min (400 R γ rays) was intermediate between the very high mutation frequency obtained after x-ray exposure of 400 R at 90 R/min and the very low frequencies obtained after γ irradiation at 0.009 R/min (400 R or 258 R), which were not significantly greater than the minimum estimate for the control rate (Russell, 1963).

At doses above 100 rad the mutational yield from acute fission-neutron exposures is well below that expected on linear extrapolation from lower doses (Batchelor et al., 1967; Russell, 1965a). As Fig. 3 shows, this phenomenon resembles that found with much higher acute x ray exposures, where the mutation frequency after a single exposure of 1000 R is lower than that after 600 R (Russell et al., 1958). Russell (1965a) has suggested that both trends result from the selective elimination at high doses of the more mutationally sensitive cells. This phenomenon has not been found after chronic neutron or γ exposures (Fig. 4).

RBEs for the induction of dominant visible mutations in mouse spermatogonia seem very similar to those for specific-locus mutations, since a comparison of chronic fission-neutron and gamma exposures has yielded a ratio of 19:1 (Batchelor et al., 1966, 1967; Searle, 1967).

The genetic results of irradiating adult female mice are shown in Table 12. A high yield of mutations is obtained with acute irradiation of the more mature oocytes, i.e. those which will be ovulated within 7 weeks of treatment. By comparison of the mutation frequency of 12.3 $\times$ 10^{-5} with 55 rad neutrons $+$ 8 rad γ rays at 79 rad/min (Russell, 1965b) with that of 19.3 $\times$ 10^{-5} reported by Russell (1963) after 400 R from x rays at 90 R/min an RBE of about 4.5 is obtained. Double this value is found if the lower frequency at 0.2 rad/min from neutrons is compared with Russell's (1963) figure for γ irradiation at 0.8 R/min. However, no specific-locus mutations were recovered from 120 483 offspring conceived more than 7 weeks after fission-neutron irradiation at high to moderate intensity (Russell, 1965b). This was in marked contrast to previous findings in males, that length of time after irradiation did not affect the mutation frequency in spermatogonia (Russell et al., 1958a). Chronic irradiation gave very similar results (Table 12), since only 1 specific locus mutation was recovered from 32 221 offspring, and this occurred in a first litter, i.e. very soon after the 12-week irradiation period (Batchelor et al., 1968; Searle and Phillips, 1968). Russell (1967) has shown that the same difference in the frequencies of mutations recovered from oocytes irradiated in earlier and later follicle stages also occurs with acute x irradiation; again there were no mutations in the over-7-week group. There are various possible explanations for this striking dependence on oocyte stage; some aspect of a genetic repair mechanism may well be involved.

Specific-locus and dominant visible mutation frequencies have been compared (Searle and Phillips, 1971) after week-long fission-neutron irradiation of early embryonic germ cells of both sexes, namely the mitotically-dividing primordial oogonia and spermatogonia and their precursors. Mutation frequencies were similar in both sexes, being 5.3 $\times$ 10^{-5}/locus in males and 6.4 $\times$ 10^{-5}/locus in females after a neutron dose of 108.5 rad delivered at 0.01 rad/min. It seems probable that the dose reaching the embryonic germ cells was about 75% of this. There are no comparable data from low-LET irradiation which can be used for RBE estimation, but the neutron irradiation was about

TABLE 12
SPECIFIC-LOCUS MUTATION FREQUENCIES AFTER ACUTE AND CHRONIC FISSION-NEUTRON IRRADIATION OF
MOUSE DICTYATE OOCYTES

| Dose in rad | | Neutron dose rate (rad/min) | Irradiation to conception interval | Mutation frequency × 10⁵ | RBE | References |
Neutron	Gamma					
55	8	79	up to 7 weeks	12.3	4.5*	
55	8	79	7 weeks	0	—	Russell (1965b)
55	8	0.2	up to 7 weeks	6.8	9.0†	
55	8	0.2	7 weeks	0	—	
80	58	0.001	0–12 months	0.4	—‡	Batchelor et al. (1968)

* Compared to 400 R x irradiation at 90 R/min (Russell, 1963).
† Compared to 400 R γ irradiation at 0.8 R/min (Russell, 1963).
‡ Comparable γ ray yield was zero (Batchelor et al., 1968).

10 times as effective as chronic γ irradiation of adult spermatogonia if dose attenuation is allowed for. It was concluded that comparable male and female germ cells are of similar genetic radiosensitivity and that the dictyate oocyte in an immature follicle is probably the only germ-cell stage in the mouse which does not yield specific locus mutations after irradiation. The human oocyte does not pass through a dictyate stage (Baker and Franchi, 1967; Oakberg and Clark, 1964), which increases the difficulty of assessing the magnitude of genetic hazards arising from the exposures of females to radiation on the basis of mouse data.

The induction of reciprocal translocations in mouse spermatogonia by fission neutrons has been studied by Searle et al. (1969) who looked for multivalent configurations in spermatocytes of the irradiated mice themselves. With acute irradiation (49–55 rad/min) the dose–response curve was similar to that for the induction of specific locus mutation, since the translocation frequency sharply decreased at high doses from a maximum of 8.7% at 100 rad to only 1.6% at 220 rad. With high chronic exposures no decrease in yield was found; in fact the translocation frequency of 21.7% after a 214 rad neutron dose was higher than expected with a linear response. If one assumes that the response is linear at low doses for both acute and chronic exposures then a comparison of the neutron data with those for low-LET radiation (Evans et al., 1970; Searle et al., 1968) leads to RBE estimates of 4 for the acute neutron vs. x-ray comparison and of about 23 for the chronic neutron vs. γ-ray comparison. The latter comparison relates to a γ-ray dose rate of 0.02 R/min; it is not known whether higher RBEs apply when the dose rate is reduced still further.

In good agreement with the above results are those of Domshlak et al. (1970) and Pomerantzeva and Ramaiya (1969), who found that the dose–effect curve for the induction of translocations by fission neutrons in mouse spermatogonia was linear over the range 18–72 rad, but fell away at higher doses. The RBE for fission neutrons vs. x rays was estimated as 4.6, while that for intermediate neutrons was 6 (Pomerantzeva and Ramaiya, 1969).

From results of studies by Ford et al. (1969) on the genetic transmission of translocations it can be estimated that the frequency of translocation heterozygotes in the offspring of irradiated males will be about one-eighth of their frequency in spermatocytes, with spermatogonial irradiation. In addition, such translocations give rise to unbalanced gametes, which act like dominant lethals in causing embryonic death around the time of implantation. The expected frequency of such dominant lethals is about one-quarter the frequency of translocations in spermatocytes. It is becoming increasingly clear (Ford et al., 1969; Lyon et al., 1964) that most of the dominant lethality resulting from spermatogonial irradiation is of this secondary type, because of selective elimination of primary dominant lethality during gametogenesis. Thus one would expect that the RBEs for induction of these secondary dominant lethals would be the same as for the induction of translocations. There is little direct evidence on this; but Batchelor et al. (1968) found a significant 4.3% decrease in the litter-size at birth of offspring of male mice given 214 rad neutrons + 93 rad γ rays over 12 weeks, compared with those given 606 rad γ rays + 2.5 rad neutrons over the same period. This is in good agreement with the decrease of 5% which would be expected on the basis of translocation results, if all dominant lethality was of the secondary type.

Translocations are also induced by fission-neutron irradiation of primordial spermatogonia in early embryos (Searle and Phillips, 1971), though the mean frequency (1.2% after 108.5 rad spread over 1 week) is not so high as when spermatogonia of adult males are irradiated. Nevertheless, the neutrons appear to be about 6 times as effective under these conditions as chronic γ irradiation of adult spermatogonia, when attenuation of the neutron dose by the tissue of the pregnant female mouse is allowed for.

Recent studies on the induction of dominant lethals in post-meiotic germ cells of male mice have, on the whole, given similar results to the earlier studies described in the RBE Report (1963). Pomerantzeva and Ramaiya (1963, 1965) reported that fission neutrons were 5–6 times as effective as γ rays and almost 4 times as effective as x rays, all given at high intensity, in inducing dominant lethal mutations in mouse spermatozoa. Spermatids were 2.5 times more radiosensitive than spermatozoa. Searle and

Phillips (1964) found that irradiation with fission neutrons at lower intensities (0.1 rad/min) was about 6 times as effective as acute x irradiation for dominant lethal induction. All these findings are in line with the earlier results of Russell and co-workers (see RBE Report, 1963) on dominant lethal induction by cyclotron neutrons (modal energy 1 MeV) and neutrons from a nuclear detonation, in which the neutron RBE was around 6 relative to x rays. Also in line with Russell's earlier findings is Tutikawa's (1967) report that 14.1 MeV neutrons were about 1.8 times as effective as acute x irradiation for the induction of dominant lethals in mouse spermatozoa.

The induction of chromosomal aberrations by high-LET irradiation of female mice has been little studied yet. Searle (1970) has irradiated female mice with fission neutrons at high intensity from the Harwell reactor BEPO and found that the dose for 50% dominant lethality is 200 rad in maturing dictyate oocytes. Since it is around 600 R for acute x irradiation of this stage (Edwards and Searle, 1963; Russell and Russell, 1956; Russell et al., 1958b) the RBE for neutrons vs. x rays is about 3.

In conclusion, it can be seen that, for a wide variety of different types of genetic damage to male germ cells, fission neutrons have an RBE of about 6 relative to x rays, while in spermatogonia this increases to around 20 if the comparison is the chronic γ radiation. There is evidence that the RBE varies with LET in the same way as in other organisms. RBE values obtained so far from fission-neutron irradiation of female germ cells at high or medium dose rates seem broadly in line with those for male germ cells, but mutational yields after protracted irradiation of mouse dictyate oocytes are so low that no RBE values can be calculated.

MAN

Doubtless radiation induces heritable mutations in man as in other organisms, but at present there seems to be no direct and unequivocal evidence of this phenomenon in human germ cells. There is, however, positive evidence of the ability of both high- and low-LET radiations to induce chromosomal aberrations in human lymphocytes. Although RBEs can only be derived from these cytogenetic data, a brief account of the genetic findings may help to put into perspective the results from experimental organisms.

An intensive search for genetic effects has been made on the offspring of survivors of the atomic bombings at Hiroshima and Nagasaki by the Atomic Bomb Casualty Commission (ABCC). The radiation emitted by the atomic bombs was mainly a mixture of fission neutrons and high-energy γ rays, with the latter predominating at Nagasaki (Auxier et al., 1966). The ABCC procedure and findings for the 1948–53 period were reported by Neel and Schull (1956). They did not find any clear-cut overall effects of maternal or paternal exposure on the frequency of congenitally malformed infants or of stillbirths or neonatal deaths, nor on birthweight or anthropometric measurements at 8–10 months of age. They concluded that "under circumstances where, on the basis of what is known concerning the radiation genetics of mammals, it appeared unlikely that conspicuous genetic effects of the atomic bombs could be demonstrated, such effects have in fact not been demonstrated". There was some evidence for an effect of exposure on the sex ratio of offspring (Neel and Schull, 1956; Schull and Neel, 1958), but later analysis did not confirm this (see Neel, 1963).

De Bellefeuille (1961) reanalysed parts of the ABCC data presented by Neel and Schull, as well as other results, and claimed that they did demonstrate harmful genetic effects of the radiation. In their rejoinder, however, Neel and Schull (1962) pointed out various flaws in De Bellefeuille's argument which rendered his conclusions invalid.

Schull et al. (1966) analysed the sex-ratio trends further by studying a group of 47 624 babies born in 1956–62. They concluded that this later material did not corroborate the suggestion of an effect of exposure on sex ratio

which had appeared in the earlier data. This suggested either that a small early effect disappeared in later births, or that the original observations had no biological significance. In connection with the former possibility, they pointed out that Russell (1965b) had found a very marked effect of time after irradiation on the frequency of mutations induced in female mice. However, they failed to find any clear evidence for a time trend in their data.

Kato et al. (1966) studied children of those exposed to the atomic bombings. They did not find significant variation in mortality ascribable to differences in parental radiation exposure. The number of years during which the children were at risk of death varied from 3 to 15, with a mean of 9. The authors used these data, in conjunction with other previously published data from Hiroshima and Nagasaki, to calculate a minimal doubling dose for man. They considered that it was unlikely that the figure was less than 50 rad for the type of radiation emitted by the atomic bombs and for mutations of the type resulting in early post-natal (i.e. juvenile) death in the first post-bomb generation. They also suggested that under the more usual conditions of human exposure to radiation the minimal doubling dose for mutations of this type was of the order of 100 rad.

The magnitude of the doubling dose for mutations causing juvenile mortality is not known in the mouse, but recent doubling-dose estimates for the induction of various other types of mutation by acute x irradiation of spermatogonia cluster around 30 R (Lüning and Searle, 1971). So far, there is very little evidence for the action of radiation-induced lethal mutations at the juvenile stage in the mouse. For instance, Lyon et al. (1964) found that 600 R + 600 R of spermatogonial x irradiation (with 8 weeks between doses) induced 10.6% dominant lethality acting in utero, but there was no evidence for any increased mortality in live-born offspring between birth and weaning.

Lüning and Sheridan (1964, 1966, 1968) have

found very little evidence for the presence of any dominant deleterious effects of recessive lethals in mice, such as might act shortly after birth. In fact, it can be said in general that genetic effects of radiation on those traits studied in the offspring of atom-bomb survivors (post-natal mortality, sex ratio, stillbirth-frequency, etc.) have not been demonstrated unequivocally in mice either, despite much higher doses. Moreoever, a number of those dominant characters in which the induction of mutations by irradiation has been clearly demonstrated in the mouse (e.g. prenatal mortality, reduced fertility due to translocation heterozygosis and internal effects on the skeleton) are very difficult to study in man. Thus there is still a dearth of valid points of comparison between the two species with respect to mutational radiosensitivity.

Schull and Neel (1962) found no increased frequency of mongolism (Down's syndrome) in the offspring of those exposed at Hiroshima and Nagasaki. In addition, Grahn and Kratchman (1963) have concluded that the excess neonatal death rate in mountainous regions of the Western United States was most probably due to a depression of foetal growth resulting from the reduced partial pressure of oxygen, rather than to extra radiation-induced mutation or foetal injury connected with the increased cosmic radiation intensity or the known uranium reserves in these regions.

Information on the RBE for high-LET radiations with respect to chromosomal abnormalities in human peripheral lymphocytes has been reviewed in a U.N. Report (1969) and by Bender (1969). Gooch et al. (1964) obtained an RBE of 2 for 14.5 MeV neutrons and of 4–5 for 2.5 MeV neutrons, compared with 250 kVp x rays, for the induction of chromosome breaks in these cells. Scott et al. (1967) exposed whole blood to fission neutrons (0.7 MeV), at both high and low intensities, and scored dicentrics and rings. There was no difference in the effectiveness of the two intensity levels, and comparison with 250 kVp x rays gave an RBE value of around 3. It is clear that the general

picture is similar to that found with induction of chromosomal aberrations in mouse germ cells.

The U.N. Report (1969) discusses the many findings of chromosomal aberrations in the blood of those exposed to radiation occupationally, in reactor accidents or in atomic explosions (see also H. J. Evans et al., 1967); detailed consideration of these would be beyond the scope of the present report.

Interpretation of the Results

In all the eukaryotic systems reviewed above, the efficiency of production of mutations in both germinal and somatic cells increases markedly with increasing LET, particularly in animals and higher plants. The effectiveness of production of chromosome structural changes observed at the first post-radiation mitosis also increases strongly with LET. It has been shown that production of these chromosome aberrations is explicable on the hypothesis that all visible chromosome aberrations arise from the interaction of *two* damaged regions of a chromosome, and that any *one* point of damage results from one energy-loss event in the DNA at that point (Neary, 1965). Where it is possible to distinguish between genetic effects due to point mutations and chromosomal structural changes such as deletions, namely in fungi, the RBE for the point mutations is considerably less than for deletions (de Serres, 1970).

On the other hand, there are strong indications that for single base-change mutations in bacteria the effectiveness of production steadily *falls* with increase of LET, in agreement with the simple picture expected if the primary lesion were due to a single energy-loss event in a molecular structure such as DNA (Bridges and Munson, 1968; Munson and Bridges, 1969). Furthermore, the absolute level of induced single base-change mutation in bacteria is between 3×10^{-11} and 1×10^{-11} per rad (Munson and Bridges, 1970); for a single base-change mutation in T4 bacteriophage the level is 1.6×10^{-11} per rad (Bridges et al., 1970).

The absolute mutation rates in the eukaryotic systems are much larger, some 10 to 100 times in fungi and up to 10 000 times in the germ cells of insects and mice, suggesting a much larger target than a single base-pair. The LET dependence and the dose-rate dependence of specific-locus mutations in mammals are similar to those for visible chromosome aberrations in a variety of cells, although there are differences in the actual patterns of dependence. These similarities might suggest that specific-locus and other similar induced mutations in mammals represented chromosome structural changes. However, W. L. Russell (1965c, 1967) has presented evidence in favour of the view that most of the mutations scored by the specific-locus method are not two-hit aberrations. He considers that the dose-rate effect for these mutations is connected with more efficient repair of single-hit mutational or premutational damage at low than at high dose rates rather than with the lower opportunity for interaction of lesions at low dose rates. There is still considerable uncertainty on the qualitative nature of radiation-induced specific-locus mutations. However, L. B. Russell (1971) has recently shown, for mutations at the closely linked *d* and *se* loci in the mouse, that only 13.5% of those recovered from unfractionated x or from γ irradiation and 31.7% of those recovered from fission-neutron irradiation of spermatogonia involved more than one functional unit. Thus the genetic damage is restricted in its chromosomal extent, though less so after fission-neutron than after x irradiation.

In general, the absolute level of mutagenic effect, its LET dependence and its dose-rate dependence must all be influenced by the capacity of the cell to repair primary radiation damage. At present, not enough is known about the nature and mechanism of the repair systems in eukaryotic cells, or about the arrangement and degree of redundancy of the genetic material in the chromosomes, for any more definite conclusions to be drawn on how to interpret the empirical data presented above.

Relationship between LET and RBE for Mutation Induction

Differences in the spatial distribution of energy in irradiated tissues (radiation quality) are known to influence the biological effectiveness of a unit of absorbed dose. Radiation quality can be specified in terms of the rate of energy loss of charged particles. The concept of linear energy transfer (LET), commonly expressed in units of kiloelectron volts per micrometre, has been discussed in a previous section. In spite of the recognised inadequacies of LET measurements to specify energy distribution in small targets important in biological (and particularly in genetic) damage, the concept has been of practical use in relating the distribution of absorbed radiation energy to its relative biological effectiveness (RBE Committee, 1963). RBE is normally expressed in terms of the pertinent biological effectiveness of ordinary x rays, taken as one (QF = 1, LET = 3.5 keV/μm).

In the Report of the RBE Committee, it was recommended that the term RBE as used in protection work be replaced by QF ("Quality Factor"). A table showing the LET-QF relationship, based on previous ICRP recommendations, was published by the RBE Committee and the figures shown are in the first two columns of Table 13. Recent (1961–70) results and some earlier ones which throw light on the RBE-LET relationship for mutation induction have been abstracted from previous chapters of the present report and are summarized in the remaining columns of Table 13. Two extra rows have been added for results from radiations of LET higher than 175 keV/μm. Results for 14 MeV neutrons have been placed in the 7–23 keV/μm category (see page 8). Fission-neutron data have been placed in the 23–53 keV/μm category, even though some estimates place their dose-average LET rather higher than this (e.g. 75 keV/μm according to Bewley, 1968), because it was felt that the RBE Committee (1963) intended them to be at the top end of this range, and to avoid excessive concentration of

TABLE 13

AVERAGE RBEs FOR DIFFERENT CLASSES OF HIGH-LET RADIATIONS WITH RESPECT TO MUTAGENESIS IN DIFFERENT ORGANISMS, AFTER ACUTE IRRADIATION

Mean relative biological effectiveness

LET$_\infty$ (keV/μm)	Quality factor (1963)	Fungi Specific-locus forward mutations Point	Fungi Specific-locus forward mutations Deletion	Higher plants Somatic mutations Specific-locus Arabidopsis	Higher plants Somatic mutations Specific-locus Zea	Higher plants Somatic mutations Chromosome aberration Nigella	Higher plants Germ-line mutations Chlorophyll loci	Insects Specific-locus mutations Silkworm late gonia	Insects Specific-locus mutations Dahlbominus oocytes	Insects Recessive lethals Drosophila pre-meiotic	Insects Translocations Drosophila post-meiotic	Mice Specific-locus mutations Spermatogonia	Mice Specific-locus mutations Late oocytes	Mice Translocations Spermatogonia	Mice Dominant lethals Spermatozoa	Mice Dominant lethals Late oocytes
<3.5	1															
3.5–7	1–2															
7–23	2–5	2	5.5	9(1.4–15)	49		15	2.5	1	2	1–2	6	4.5	4	2	3
23–53	5–10			16	63	66	22 (3–40)	4			4			6	6	
53–175	10–20			18	90	89	29									
175–900	—	5.5	74	11.5												
>900	—			1.5												

results in the 53–175 keV/μm category. Only data derived from acute exposures are given in the table; results of chronic exposures are discussed later. The figures abstracted for insects and mice refer particularly to effects in the germ cells that correspond to those most at risk in man (spermatogonia and oocytes), although some other data throwing light on the RBE–LET relationship have been included.

The data on mutational effects are most extensive in higher plants. They include observations made: (i) in the treated generation (R_1) on chromosomal aberrations and on somatic cell lineages marked by deletion of a specific locus: (ii) in the progeny (R_2) as measured by the frequency of chlorophyll-deficient mutations, which is the most commonly used genetic index. The range of RBEs found is so wide (see Tables 4–6) and varies so much with species, tissue irradiated and the exact conditions of the experiments, that overall means could be misleading and biased. Therefore the means and ranges given for somatic mutations refer only to the few experiments on particular species in which the RBE–LET relationship was specifically explored by using radiations of a number of different LETs. When estimates differed widely the range is given as well as the mean. Since no experiments on progeny of irradiated plants explored a range of LETs, all the RBE estimates for chlorophyll mutations have been used for deriving RBEs for genetic effects. These are, on average, 2–3 times higher than the appropriate quality factors used by ICRP. For both somatic and germ-line mutations there is a consistent increase in RBE as one proceeds from the 7–23 keV/μm LET category to the 53–175 keV/μm one. However, it can be seen from the specific locus somatic mutation results in *Arabidopsis* that at still higher LETs the RBE declines again. This suggests that if the term "high-LET radiation" is defined on the basis of radiobiological effects (see page 5) it should have an upper as well as a lower limit.

In fungi, a very high RBE of 74 was found for induction of specific locus deletions at an LET above the range given in the RBE Report, but comparable results at fission neutron LETs are not yet available.

The RBE values for insects (three separate orders) and for mice are very similar on the whole and are in good agreement with the Quality Factors recommended by the RBE Committee, especially with the lower values in the relevant ranges. Although RBE values are decidedly lower than those for somatic and germ-line mutations in plants they agree with them in showing an increase in RBE with increasing LET up to the 53–175 keV/μm range. However, the amount of information available is still very limited.

Table 13 lists results for acute exposures only, but several studies on the mutagenic effects of protracted exposures to high LET radiation have been carried out in mice. These have shown that chronic exposures to fast neutrons are about 20 times as effective as chronic γ exposures for the induction of specific locus and dominant visible mutations, as well as for the induction of reciprocal translocations. The RBE Committee discussed a similar phenomenon in higher plants, where very high RBE values were obtained when the induction of chromosome aberrations by protracted neutron and γ exposures was compared (see Neary *et al.*, 1963). High values at low doses or dose rates would be expected on theoretical grounds when the dose–response relationship for the low-LET radiation contains a quadratic (dose-squared) as well as a linear term, while that for high LET radiation contains a linear term only. RBE values would then approach a maximum as the dose or dose rate was lowered and the quadratic term in the low-LET equation tended to disappear. The RBE Committee gave the RBE value at minimal doses a special symbol RBE_M and defined it as "the ratio of slopes of the dose–effect curves at zero dose".

The increased RBE for specific-locus mutations in spermatogonia of mice after chronic exposures results from the fact that acute x irradiation is about 3.3 times as effective as chronic γ irradiation for the induction of this type of mutation (Russell, 1965a). A lower capacity for repair of pre-mutational damage after acute than after chronic irradiation is believed to be the main reason for this difference (Russell *et al.*, 1958a).

We can conclude from this synthesis that RBEs for mutagenesis tend to increase with increasing LET in the same way as those for other types of biological effect in cells of higher organisms. Moreover, in experiments with animals, RBE values for genetic effects of acute radiations of different LET are in reasonable agreement with corresponding QF values recommended by the ICRP. However, experiments involving chronic exposures of male mice have yielded RBEs which are decidedly higher than the QF value. These high RBE values result from a decreased effectiveness of the low-LET irradiation at low dose rates rather than from an increased effectiveness of the high-LET irradiation.

Factors likely to Influence the RBE–LET Relationship

Although for protection purposes it is necessary to assign a fixed Quality Factor to radiation of a particular LET or range of LETs (perhaps with the addition of modifying factors when certain organs are irradiated or in other circumstances) yet experimental work has shown that the actual RBE with respect to genetic damage does depend on a number of biological and physical factors. It seems desirable to deal with all of these together in one section, though some have already been discussed.

DOSE AND DOSE RATE

The influence of these factors was discussed at length in the report of the RBE Committee (1963) and some of the consequences were described in the last section. Dose–response curves for high-LET radiations tend to be more linear than those for low-LET radiations and are less influenced by dose rate. Therefore the RBE value depends on the radiation dose and dose rate used. It is necessary to decide the most meaningful level of response at which to derive a single RBE value from the whole range. Usually the level chosen by workers has been within the range of experimental observation rather than an extrapolation from them; thus for plants most mutations were scored at a level of 1–3% and it was at this level that RBEs were calculated (see section on higher plants).

For radiation protection purposes, it is important to know what RBE values are likely to become at low doses and dose rates. Unfortunately there are few mutagenic test systems which will work efficiently at the levels required. Therefore it has been necessary in this report to consider results from higher doses and dose rates also. These illuminate the LET vs. RBE relationship and help one to deduce the probable RBE values at actual levels of exposure. There are two main approaches to this problem. One is to assume that the high-LET response is linear and the low-LET response curvilinear at all levels of exposure, so that the RBE values continue to rise as the dose and/or dose rate continues to fall. The other is to assume that, because of a linear term in the low-LET dose–response equation, the low-LET response gradually changes from curvilinear to linear, so that the RBE value reaches a maximum and then remains constant. There are good grounds for preferring the second model when dealing with structural changes involving interaction between lesions (Neary *et al.*, 1963). The best approach is less obvious, however, when the "single-hit" type of event is being considered where the greater effectiveness of high-LET radiation may well be connected with the extent of damage to a quite different entity, for example, the repair mechanism of the cell. However, W. L. Russell (1963, 1965a) reported that although yields of specific locus mutations from spermatogonial irradiation of mice declined considerably as the dose rate was lowered from 90 R/min to 0.8 R/min, there was little evidence for any further decrease between 0.8 R/min and 0.001 R/min. Since the yield from fission neutrons was little affected by dose-rate change (Batchelor *et al.*,

1966; Russell, 1965a) at lower doses, these observations also favour the second idea of a constant maximum RBE at low dose rates, and presumably at low doses also. Thus, Sparrow *et al.* (1972) found that RBE values for induction of pink cells (a probable deletion event) in *Tradescantia* stamen hairs by 0.43 monoenergetic neutrons vs. 250 kVp x rays continued to increase down to about 0.1 rad neutrons and 5 rad x rays. Below this dose the x ray curve becomes linear and therefore parallel to the neutron curve, so that the RBE reaches a constant value of about 50 at lower doses.

In mice, the relationship between the frequency of translocations observed in spermatocytes after spermatogonial exposure and the dose of acute x or γ irradiation has agreed with a linear hypothesis between doses of 25 and 800 rad (Léonard and Deknudt, 1967, 1968; Evans *et al.*, 1970). At higher doses of x rays and at fission neutron doses above 100 rad, however, there is a marked falling off in yield (Lyon and Morris, 1969; Searle *et al.*, 1969), presumably as a result of germinal selection. The true undistorted response for low-LET irradiation is likely to be curvilinear (with a linear and a quadratic component), so some increases in RBE values would be expected at very low doses. This is already known to occur at low dose rates (Searle *et al.*, 1969). With specific-locus mutations also, an increase in RBE seems likely at low doses as well as at low dose rates (Batchelor *et al.*, 1965; Russell, 1965a) since Russell and Kelly (1965) found that the yield of specific-locus mutations from exposure of maturing dictyate oocytes to 50 R acute x rays was significantly lower than expected from data at 400 R. Similar data from spermatogonial irradiation are not yet available. Probably the RBE values at very low dose rates are the best indication of their likely magnitude at very low doses.

Surprisingly, there is evidence for non-linearity of translocation yield after chronic neutron exposures of mouse spermatogonia (Searle *et al.*, 1969); the yield at the lower neutron dose was used for the RBE calculation. There is also some evidence (Russell, 1965b) that in female mice a

lowered dose rate of fission neutrons induces fewer specific-locus mutations, even when all the dose goes to the genetically radiosensitive mature stage. The exceptionally low yield after protracted exposures to fission neutrons seems to be a special phenomenon connected with the remarkable properties of the immature dictyate oocyte in mice, from which no specific-locus mutations have yet been found by Russell (1965b, 1967) in a very large number of progeny.

Oftedal (1964) found evidence for a greater efficiency of low than higher doses of low-LET radiation for the induction of sex-linked recessive lethals in *Drosophila* spermatogonia, which he attributed to the survival of a class of highly sensitive cells at the lower dose. However, no signs of a similar phenomenon in mouse spermatogonia have been found when low doses were used (Carter *et al.*, 1958; Léonard and Deknudt, 1968). Good evidence for major effects of germ cell killing in mice has only been obtained at very high doses, well above the level of interest for genetic aspects of radiation protection.

DOSE FRACTIONATION

By analogy with the results of experiments on somatic damage, one might expect that mutational yields would be less affected by fractionation of high-LET exposures than of low-LET ones. However, the genetic effects of fractionating radiation of high LET have as yet been little studied, and there is no evidence at present for substantial differences in response from those of low-LET radiation. Thus Murakami *et al.* (1966) found that fractionating a dose of 14 MeV neutrons could more than double the frequency of specific-locus mutations resulting from irradiated silkworm spermatogonia, with a similar enhancement effect after low-LET irradiation (x and γ rays), except that the peak yield occurred at an 18-hr interval between doses with low-LET and a 36-hr interval with high-LET irradiation. Both findings may have resulted from partial synchronization of germ-cell stages following the first dose, so that a sensitive stage was reached at the time of the second dose.

Fractionation with short intervals between doses of low-LET radiation frequently leads to a lower than expected yield; for example, Lyon *et al.* (1970) found that when a dose of 300 rad x rays was delivered to mouse spermatogonia in successive daily doses the resultant yield of translocations was markedly less than when a single 300 rad dose was given, although the dose–response relationship for translocation induction in mouse spermatogonia by single doses (but observed in spermatocytes) is effectively linear. Similar experiments with high-LET radiation have not yet been carried out. When there is a long interval between fractions an additive response has been found with both high-LET (Searle *et al.*, 1971) and low-LET radiations (Russell, 1963).

SEX AND TYPE OF GERM CELL

Although most of the dose of high-LET radiations received by exposed persons during their reproductive life will be absorbed by germ cells when they are spermatogonia (in males) or resting oocytes (in females) yet it is of some importance to know whether markedly different RBE values are associated with any other types of germ cell. The evidence so far suggests that although the genetic radiosensitivities of pre- and post-meiotic germ cells may differ greatly, RBEs for mutation induction do not show much change. Thus Lamb *et al.* (1967) found very similar RBE values for recessive lethal induction with fast neutrons in pre- and post-meiotic germ cells of *Drosophila*. In the mouse, the RBE for dominant lethal induction in mouse spermatozoa by fast neutrons is around 6, while that for translocation-induction in spermatogonia (which leads to dominant lethality as a secondary consequence) is about 4 for similar acute exposures (see section on mammals). For maturing dictyate oocytes a value of 3 has been obtained for dominant-lethal induction (fast neutrons vs. x rays). Among post-meiotic germ cells in the male some changes in RBE values

might be expected because the differing radiosensitivities (of spermatozoa and spermatids for example) may depend partly on oxygen-tension differences, which affect responses to low-LET radiation much more than to high. Thus Sobels and Broerse (1970) considered that the differences they found between late spermatids and mature spermatozoa in RBE values for induction of various types of mutational event in *Drosophila* were connected with the lower degree of oxygenation in late spermatids than in sperm, leading to a higher effectiveness of fast neutrons in the former. Possibly the same phenomenon occurs in mammals, but the overall picture would be affected little.

For both high- and low-LET irradiation the genetic radiosensitivities of embryonic or foetal germ cells seem on the whole to be somewhat lower than in the adult (see Searle and Phillips, 1971). Therefore, although no experiments directly comparing genetic effects of high- and low-LET irradiation have been carried out in these germ cells it seems probable that RBEs would differ little from those in the adult.

Results of embryonic high-LET irradiation by Searle and Phillips (1971) suggest that there is no intrinsic difference between genetic radiosensitivities of male and female germ cells. However, as already stated, the mouse dictyate oocyte when in immature follicles shows a remarkably low genetic radiosensitivity with both high- and low-LET radiation (Russell, 1965b, 1967) so that no RBE value can be calculated. It is not known to what extent this finding applies to the immature human oocyte. RBE values for the mouse dictyate oocyte in maturing follicles are, however, similar to those for spermatogonia (see section on mammals).

OXYGEN

For most radiobiological effects it has been found that the enhancing effect of the presence of oxygen at the instant of irradiation becomes less pronounced as the LET is increased. Thus the RBE–LET relationship may be appreciably different for irradiations under fully oxygenated and oxygen-depleted conditions. The influence of oxygen at various defined LET values on the magnitude of mutagenic effect appears to have been investigated systematically only for yeast (Manney et al., 1963; Mortimer et al., 1965) and here it was found that as in other systems the oxygen enhancement ratio (OER) decreased progressively as the LET increased. At values of LET above about 200 keV/μm, however, the OER increased again somewhat, an effect attributed to a greater importance of the δ rays relative to the track core which at these values of LET is beginning to be inefficient. The energies of the δ rays of a heavy ion with an energy of 10 MeV per atomic mass unit extend up to 22 keV and the LET spectrum falls in a relatively low region so that the OERs of the δ rays are quite considerable. The overall OER may thus be appreciably greater than 1.

The RBE for mutagenic effects in yeast increases with LET rather less than in other systems; hence the yeast data on the influence of oxygen are unlikely to be directly applicable to these other systems. The other radiobiological systems for which the effect of oxygen has been systematically measured at various LET values are mammalian cell killing (Barendsen, 1968; Berry, 1970; Bewley, 1968; Todd, 1967), yeast cell killing (Manney et al., 1963; Mortimer et al., 1965), bacterial cell killing (Alper et al., 1967; Deering, 1963), bacterial spore inactivation (Powers et al., 1968) and visible chromosome aberrations (Neary et al., 1967). There is no identity between the effects of oxygen in any of these systems, except that all demonstrate the general fall in OER with increasing LET.

It follows that the precise effect of oxygen on the RBE–LET relationship for mutagenic effects in mammals cannot be inferred from data on other systems at present. However, those germ cells in mammals which are most at risk of radiation exposure seem to be quite well oxygenated, judging from results of oxygen tension experiments on mouse spermatogonia (Ashwood-Smith et al., 1965; Hornsey et al., 1971) and the state of vascularization of the mammalian ovary (Harrison, 1962). Therefore,

this can be taken as the relevant condition for the RBE–LET relationship. Thus, except in unlikely special circumstances, variability in the relationship for mutagenic effects due to changes in oxygenation is unlikely to arise in practice.

TYPE OF MUTATION

There is some evidence that there may be a qualitative difference between mutations induced by high- and low-LET radiation. In Neurospora, de Serres (1970) found that higher-LET radiation is more efficient in the induction of recessive lethal mutations by chromosome deletion than by point mutation. In the mouse, there is evidence that more d-se deficiencies are induced by fission neutrons than by x or γ radiation, although the frequency of homozygous lethality among specific-locus mutations induced by the two types of radiation seems very similar (Batchelor et al., 1966; Russell, 1965; Russell and Russell, 1959). In barley it has been found (see section on plants) that 30% of the eceriferum mutants induced by high-LET radiations were at one particular locus while those induced by low-LET radiations were distributed rather evenly among a large number of loci. On the other hand, in the mouse, the spectrum of specific-locus mutations (apart from the d-se mutation) in the original 7-locus stock is very similar after low- and high-LET irradiation, with the s locus predominating in each case (Batchelor et al., 1966; Russell and Russell, 1959). Neither are there any obvious qualitative differences between the translocations induced by fast neutrons and x rays in the mouse. On present evidence, the types of difference found are unlikely to have much effect on the extent to which the mutations induced are deleterious.

Applicability to Man

Table 13 shows that although there are large differences between RBE values for plants and animals, those found in insects and in mice show fairly close agreement, although comparisons are only possible over a limited range of

LETs. For acute exposures, therefore, it seems fairly safe to assume that RBE values for man are of the same order. There seems no doubt that the general form of RBE–LET relationship holds for human as well as animal cells. This is shown, for example, by the work of Barendsen et al. (1966) on survival in cultured human cells after exposure to radiations giving a wide range of LETs. The D_{37} values declined from a maximum with 250 kVp x rays (averaged LET = 1.3 keV/μm) to a minimum with 4.0 MeV α-particles (LET = 110 keV/μm) and cells in equilibrium with air, or 3.4 MeV α-particles (LET = 140 keV/μm) when cells were in equilibrium with nitrogen.

With chronic exposures, RBE values for gene mutations in spermatogonia are presumably higher in mice than would be expected in Drosophila, because of the pronounced dose-rate effect in the former (Russell, 1963) and its apparent absence in the latter (Muller et al., 1963; Purdom, 1963). However, no really comparable studies have been made from this point of view. As already mentioned, it has been suggested that the dose-rate effect in mice is connected with the reparability of premutational lesions. Repair systems of the same type as are found in micro-organisms have already been discovered in human cells (Cleaver, 1969).

In summary, none of the factors discussed seem likely to make overall RBE values for mutation induction in man after high-LET exposures very different from those already reported in the mouse and other experimental animals.

Implications for Radiation Protection

We have found that RBEs for radiation-induced genetic damage, as for somatic, tend to rise with increasing LET, whatever the type of mutation being studied. For acute irradiation of animals the genetic RBEs reported usually lie near the lower end of the range of QF values suggested by the RBE Committee for each range of LETs adopted by the ICRP (1966). From the limited amount of work on chronic exposures,

RBEs have been reported with decidedly higher values than the appropriate QF.

In considering the implications of these findings for radiation protection it is first necessary to discuss the bases of the Commission's recommendations on maximum permissible doses for genetic effects. Although the need to minimize genetic effects by limiting the exposure, both to individuals and to the population as a whole, is stressed in both the 1959 and 1966 ICRP recommendations, the maximum permissible genetic dose is explicitly discussed only with reference to population exposure. In its 1959 publication the Commission gave some explanation of how the genetic dose limit of 5 rem plus the dose from medical exposures was arrived at (ICRP Publication 1, paragraph 19). It was apparently derived from estimates "made by different national and international scientific bodies" that a gonad dose of 6–10 rem, accumulated from conception to age 30, would add a considerable social burden because of extra genetic damage, but that this could be considered tolerable and justifiable because of the benefits accruing from the practical applications of atomic energy. The reports referred to no doubt include those of the United Nations (1958), U.S. National Academy of Sciences (1956), U.K. Medical Research Council (1956). The evidence from experimental work which the authors of these reports considered was mainly based on the results of acute irradiation, since little information on the induction of mutations by chronic irradiation had yet been published. Thus the lowered yield of specific-locus mutations after chronic irradiation of mice (Russell et al., 1958a) had not been announced at the time these reports were prepared. The influence of dose rate was, in fact, discussed in the ICRP's 1966 Recommendations where it is stated (paragraph 23) that "The Commission's previous recommendations have been based on numerical considerations derived from genetic experiments using high dose rates and may seem somewhat conservative in the light of these newer experiments." However, the Commission did not think it advisable to modify its recommendations to allow for a probable influence of dose rate.

Since present recommendations for genetic effects are based on data from acute irradiation at fairly high levels of exposure it is clear that experimental RBE values under such conditions are the most relevant ones to consider when discussing the most appropriate QF figures. As stated earlier, RBEs found in animal experiments for germ cells most at risk are in good agreement with QF values recommended by the Commission. Studies in other organisms indicate that the general RBE–LET relationship found for somatic damage holds also for mutational damage. Thus there seems to be no need for any different set of values for genetic effects. The situation would have to be reviewed however, if the Commission were to revise its recommendations and base them instead on mutation induced by protracted low-LET irradiation. Much higher RBEs would then apply, as shown by the experimental data on the induction of specific-locus mutations and translocations in mice after chronic exposures, described earlier. There would then be a strong case for a substantial increase in the QF values applicable for protection from genetic effects. In the meantime it is important to remember that chronic exposures to high-LET radiation can be as genetically effective as acute exposures, thus fully justifying the continuance of safety precautions, with associated shielding and monitoring devices, wherever these are deemed necessary because of fast neutron or other hazards.

Conclusions

There is little to add to the remarks in the last section. The length of this report shows how much information has become available since the RBE Report (1963), although there are still several important gaps. In particular, it would be useful to have genetic information from animals irradiated with a wider range of LETs, and from mammals with oocytes more comparable to the human one than is the mouse oocyte. However, the general RBE–LET relationship has been very clearly demonstrated by the work on plants, even though (for some unknown reason) RBE values tend to be much higher than in animals. The reasonable correspondence between those for insects and for mice allows one to predict with some confidence that similar values would be found in man and that therefore the present QF values are appropriate. We have not dealt with the hypothetical question of what would be the appropriate QF values for high-LET radiations if maximum permissible doses for genetic effects were to be based on expectations at low doses and dose rates, but they would presumably have to be increased to about the same extent as the maximum permissible levels of low-LET radiations.

Summary

1. The RBE for neutrons and other high-LET radiations with respect to mutagenesis is reassessed in the light of extensive information which has accrued since the RBE Report (1963).

2. Available information on present-day exposures to high-LET radiation from galactic, solar and artificial sources is reviewed. The levels of population and occupational exposure are generally very low though liable to rise somewhat in the future with increased supersonic travel and use of nuclear fuel.

3. The characteristics of high-LET radiations used experimentally are described, as well as the problems associated with the LET concept. The advantages of using dose-average LET, where there is a distribution of radiation quality, are pointed out. Reasons why this measure does not reflect the biological effectiveness of 14 MeV neutrons are discussed.

4. Experimental work with high-LET radiations on micro-organisms, fungi, higher plants, insects (silkworm, *Drosophila* and some Hymenoptera) and mammals is reviewed, as well as studies on the off-

spring of atomic bomb survivors. The calculated RBE values vary greatly in different groups and at different LETs. Highest values (up to 115) were found in higher plants especially with respect to single-locus mutations and chromosome aberrations in somatic cells after acute irradiation of seeds. High values were also found after chronic exposures. Values in insects and mammals (mice) were similar and generally higher than those suggested in the RBE Report. RBEs for gene mutations (including small deletions) and for gross structural changes in animals showed no marked differences, but work on fungi showed that much higher RBEs were found with induction of deletions than with true point mutations (such as base-pair changes).

5. A synthesis of the findings for acute exposures (Table 13) and a comparison with recommended Quality Factors (QFs) shows that there is a general tendency for RBE values to increase with LET up to and including the 53–175 keV/μm range, but the limited data suggest that there is a fall at higher LETs. Although in higher plants RBE values tend to be much higher than recommended QFs, in insects and mammals there is general good agreement between RBE and QF values, with RBE values tending to correspond to the lower part of the QF range.

6. Actual and likely changes in RBE values with decreasing doses and dose rates are considered. There is ample evidence that RBE values rise under these circumstances (for example from 6 to 20 with specific-locus mutations in mouse spermatogonia, when the effects of fission neutrons are compared with those of chronic exposures to low-LET radiations rather than acute exposures). It was felt that maximum RBE values would be reached at low levels of occupational or population exposure. These maximum values would result from a decreasing effectiveness of the low-LET irradiation rather than an increased effectiveness of the high.

7. Other factors likely to influence the RBE–LET relationship in man are discussed (e.g. sex and type of germ cell, oxygen tension) but none seem likely to make RBE values for mutagenesis in man very different from those in the mouse and other experimental animals.

8. Implications of these findings for radiation protection are discussed. It is clear that the ICRP's present recommendations on maximum permissible doses for genetic effects are based on results of fairly high acute exposures. Therefore the RBE results for acute exposures apply and there is no need for any change in recommended QF values. However, if the Commission's recommendations were to be based instead on probable genetic effects at low doses and dose rates, then much higher RBE values would apply and QF values would have to be revised.

9. The Task Group stress that fission neutrons and other high-LET radiations are known to induce gene mutations and chromosome aberrations even at very low doses and dose rates. Therefore continuance of the present safety precautions, designed to reduce occupational exposures to very low levels, is fully justified.

References

ALPER, T., MOORE, J. L. and BEWLEY, D. K. (1967) LET as a determinant of bacterial radiosensitivity, and its modification by anoxia and glycerol. *Radiat. Res.* **32**, 277–293.

ASHWOOD-SMITH, M. J., EVANS, E. P. and SEARLE, A. G. (1965) The effect of hypothermia on the induction of chromosome mutations by acute X-irradiation by mice. *Mutat. Res.* **2**, 544–551.

AUXIER, J. A., CHEKA, J. S., HAYWOOD, F. F., JONES, T. O. and THORNGATE, J. H. (1966) Free-field radiation-dose distributions from the Hiroshima and Nagasaki bombings. *Hlth Physics* **12**, 425–429.

AUXIER, J. A., JONES, T. D. and HUBBELL, H. H. (1969) Review of depth-dose calculation and experimentation. *Symposium on Neutrons in Radiobiology, Oak Ridge, Tenn.* CONF-691106, 73–90.

BAKER, T. G. and FRANCHI, L. L. (1967) The fine structure of oogonia and oocytes in human ovaries. *J. Cell. Sci.* **2**, 213–224.

BALDWIN, W. F. (1968) Increased yield of gamma-induced eye colour mutations from chronic versus acute exposures in *Dahlbominus. Isotopes and Radiations in Entomology.* Vienna, International Atomic Energy Agency, pp. 365–375.

BALDWIN, W. F. and CROSS, W. G. (1966) Effects of fast neutrons on eye colour mutations in *Dahlbominus. Nature* **210**, 1396–1397.

BARENDSEN, G. W. (1968) Responses of cultured cells, tumours and normal tissues to radiations of different linear energy transfer. *Curr. Top. Radiat. Res.* **4**, 243–356.

BARENDSEN, G. W., KOOT, C. J., VAN KERSEN, G. R., BEWLEY, D. K., FIELD, S. B. and PARNELL, C. J. (1966) The effect of oxygen on impairment of the proliferative capacity of human cells in culture by ionizing radiations of different LET. *Int. J. Radiat Biol.* **10**, 317–327.

BATCHELOR, A. L., PHILLIPS, R. J. S. and SEARLE, A. G. (1966) A comparison of the mutagenic effectiveness of chronic neutron- and γ-irradiation of mouse spermatogonia. *Mutat. Res.* **3**, 219–229.

BATCHELOR, A. L., PHILLIPS, R. J. S. and SEARLE, A. G. (1967) The reversed dose-rate effect with fast neutron irradiation of mouse spermatogonia. *Mutat. Res.* **4**, 229–231.

BATCHELOR, A. L., PHILLIPS, R. J. S. and SEARLE, A. G. (1968) The ineffectiveness of chronic irradiation with neutrons and gamma rays in inducing mutations in female mice. *Br. J. Radiol.* **42**, 448–451.

BENDER, M. A. (1969) Human radiation cytogenetics. *Adv. Radiat. Biol.* **3**, 215–275.

BERRY, R. J. (1970) Survival of murine leukaemia cells *in vivo* after irradiation *in vitro* under aerobic and hypoxic conditions with monoenergetic accelerated charged particles. *Radiat. Res.* **44**, 237–247.

BEWLEY, D. K. (1968*a*) Calculated LET distributions of fast neutrons. *Radiat. Res.* **34**, 437–445.

BEWLEY, D. K. (1968*b*) A comparison of the response of mammalian cells to fast neutrons and charged particle beams. *Radiat. Res.* **34**, 446–458.

BHATT, B. Y., BORA, K. C. and PATIL, S. H. (1961) Influence of ploidy-dose levels on RBE and the process of elimination of chromosomal aberrations. *Effects of Ionizing Radiations on Seeds.* Vienna, I.A.E.A., pp. 441–459.

BOAG, J. W. (1954) The distribution of linear energy transfer or "Ion Density" for fast neutrons in water. *Radiat. Res.* **1**, 323–341.

BOOT, S. J. and DENNIS, J. A. (1968) Flux density distributions in and around a man-sized phantom irradiated with thermal neutrons. *Phys. Med. Biol.* **13**, 573–583.

BORA, K. C. (1961) Relative biological efficiencies of ionizing radiations on the induction of cytogenetic effects in plants. *Effects of Ionizing Radiations on Seeds.* Vienna, I.A.E.A., pp. 345–357.

BRIDGES, B. A. and MUNSON, R. J. (1968) Genetic radiation damage and its repair in *Escherichia coli. Curr. Top. Radiat. Res.* **4**, 95–188.

BRIDGES, B. A., DENNIS, R. E. and MUNSON, R. J. (1970) Mutagenesis in *Escherichia coli.* III. Requirement for DNA synthesis in mutation by gamma rays of T4-phage complexed with *Escherichia coli. Genet. Res.* **15**, 147–156.

BROERTJES, C. (1968) Dose-rate effects in *Saintpaulia. Mutations in Plant Breeding II.* Vienna, I.A.E.A., p. 63.

BRUCE, W. R., PEARSON, M. L. and FREEDHOFF, H. S. (1963) The linear energy transfer distributions resulting from primary and scattered X-rays and gamma rays with primary HVL's from 1.25 mm Cu to 11 mm Pb. *Radiat. Res.* **19**, 606–620.

BRUSTAD, T. (1962) Heavy ions and some aspects of their use in molecular and cellular radiobiology. *Adv. biol. med. Phys.* **8**, 161.

CARTER, T. C., LYON, M. F. and PHILLIPS, R. J. S. (1958) Genetic hazard of ionizing radiations, *Nature,* **182**, 409.

CLEAVER, J. E. (1969) Xeroderma pigmentosum, a human disease in which an initial stage of DNA-repair is defective, *Proc. Natn. Acad. Sci. (U.S.A.)* **63**, 428–435.

COOK, J. E. (1958) Fast neutron dosimetry using nuclear emulsions. AERE HP/R 2744. AERE, Harwell, Berks., U.K.

DANCER, G. H. C. (1971) Personal communication.

DAUCH, F., APITZSCH, U., CATSCH, A. and ZIMMER, K. G. (1966) RBE schneller Neutronen bei der Auslösung von Mutationen bei *Drosophila melanogaster. Mutat. Res.* **3**, 185–193.

DAVIES, D. R. and BATEMAN, J. L. (1963) A high relative biological efficiency of 650 keV neutrons and 250 kVp x-rays in somatic mutation induction. *Nature* **200**, 485–486.

DE BELLEFEUILLE, P. (1961) Genetic hazards of radiation to man. Part I. *Acta Radiol.* **56,** 65–80.

DERRING, R. A. (1963) Mutation and killing of *Escherichia coli* WP-2 by accelerated heavy ions and other radiations. *Radiat. Res.* **19,** 169–178.

DE SERRES, F. J. (1970) Characterization of heavy ion-induced mutations in the *ad*-3 region of a two-component heterokaryon of *Neurospora crassa. Biol. Div. Ann. Progr. Rept. Dec.* 31, 1969, ORNL-4535.

DE SERRES, F. J., WEBBER, B. B. and LYMAN, J. T. (1967) Mutation-induction and nuclear inactivation in *Neurospora crassa* using radiations with different rates of energy loss. *Radiat. Res.* (Suppl. 7), 160–171.

DICKERMAN, R. C. (1967) Fast neutron and X-ray irradiation of *Drosophila melanogaster* oogonia and oocytes. *Genetics,* **56,** 555–556.

DOMSHLAK, M. G., POMERANTZEVA, M. D. and RAMAIYA, L. K. (1970) The mutagenic effect of different types of radiation on spermatogonia in mice. IV. The genetic effect of fast neutrons. *Genetika* **6,** 73–82.

DUNSTER, H. J. (1970) Personal communication.

ECOCHARD, R. (1970) An approach to the study of genetic effects from the ^{14}N(n,p)^{14}C reaction for thermal neutrons. *Int. J. Radiat. Biol.* **17,** 439–448.

EDWARDS, R. G. and SEARLE, A. G. (1963) Genetic radiosensitivity of specific post-dictyate stages in mouse oocytes. *Genet. Res.* **4,** 389–398.

EVANS, E. P., FORD, C. E., SEARLE, A. G. and WEST, B. J. (1970) Studies on the induction of translocations in mouse spermatogonia. III. Effects of X-irradiation. *Mutat. Res.* **9,** 501–506.

EVANS, H. J., COURT BROWN, W. M. and MCLEAN, A. S. (1967) *Human Radiation Cytogenetics.* Amsterdam, North-Holland Pub. Co.

FLUKE, D. J. and FORRO, F. JR. (1960) Efficiency of inactivation of dry T-1 bacteriophage by protons, deuterons, and helium ions from a 60-inch cyclotron. *Radiat. Res.* **13,** 305–317.

FORD, C. E., SEARLE, A. G., EVANS, E. P. and WEST, B. J. (1969) Differential transmission of translocations induced in spermatogonia of mice by irradiation. *Cytogenetics,* **8,** 447–470.

FRIGERIO, N. A. and BRANSON, M. H. (1969) Current research in depth dose calculations. *Symposium on Neutrons in Radiobiology, Oak Ridge, Tenn.* CONF-691106, 95–115.

FUJII, T. (1964a) Relative biological effectiveness of 14 MeV fast neutrons to ^{60}Co gamma-rays in einkorn wheat. *Biological Effects of Neutron and Proton Irradiations* 2. Vienna, I.A.E.A., pp. 217–231.

FUJII, T. (1964b) Radiation effects on *Arabidopsis thaliana*. I. Comparative efficiencies of γ-rays, fission and 14 MeV neutrons in somatic mutations. *Jap. J. Genet.* **38,** 91–101.

FUJII, T. (1969) Relative biological effectiveness of high LET radiations in higher plants. *Jap. J. Genet.* **44,** Suppl. 1, 431–442.

FUJII, T., IKENAGA, M. and LYMAN, J. T. (1966) Radiation effects on *Arabidopsis thaliana*. II. Killing and mutagenic efficiencies of heavy ionizing particles. *Radiat. Botany* **6,** 297–306.

FUJII, T., IKENAGA, M. and LYMAN, J. (1967) Killing and mutagenic efficiencies of heavy ionizing particles in *Arabidopsis thaliana. Nature* **213,** 175–176.

GOOCH, P. C., BENDER, M. A. and RANDOLPH, M. L. (1964) Chromosome aberrations induced in human somatic cells by neutrons. *Biological Effects of Neutron and Proton Irradiations.* Vienna, I.A.E.A., pp. 325–342.

GOPAL-AYENGAR, A. G. and SWAMINATHAN, M. S. (1964) Use of neutron irradiation in agriculture and applied genetics. *Biological Effects of Neutron and Proton Irradiations* 1. Vienna, I.A.E.A., pp. 409–432.

GRAHN, D. and KRATCHMAN, J. (1963) Variation in neonatal death rate and birth weight in the United States and possible relations to environmental radiation, geology and altitude. *Am. J. hum. Genet.* **15,** 329–352.

GUTHRIE, M. P., ALSMILLER, R. G. and BERTINI, H. W. (1968) Calculation of the capture of negative pions in light elements, and comparison with experiments pertaining to cancer radiotherapy. *Nucl. Instrum. Meth.* **66,** 29–36.

HAGBERG, A., GUSTAFSSON, A. and EHRENBERG, L. (1958) Sparsely contra densely ionizing radiations and the origin of erectoid mutations in barley. *Hereditas* **44,** 523–530.

HARRISON, R. J. (1962) The structure of the ovary. *The Ovary*, ed. S. ZUCKERMAN, New York and London, Academic Press, pp. 143–187.

HILL, M. J. (1971) Personal communication.

HIRONO, Y., SMITH, H. H., LYMAN, J. T., THOMPSON, K. H. and BAUM, J. (1970) Relative biological effectiveness of heavy ions in producing mutations, tumors and growth inhibition in the crucifer plant, *Arabidopsis. Radiat. Res.* **44,** 204–223.

HORNSEY, S., BRYANT, P. E. and HEDGES, M. J. (1971) The effect on the sensitivity of the mouse testis of different oxygen tensions during irradiation. *Int. J. Radiat. Biol.* **19,** 21–26.

HOWARD-FLANDERS, P. (1958) Physical and chemical mechanisms in the injury of cells by ionizing radiations. *Adv. biol. med. Phys.* **6,** 553–603.

HRISHI, N. and JAMES, A. P. (1964) The induction of mutation in yeast by thermal neutrons. *Can. J. Genet. Cytol.* **6,** 357–363.

ICRP (1959) *Recommendations*, adopted September 9, 1958. ICRP Publication 1. London, Pergamon Press.

ICRP (1966) *Recommendations*, adopted September 17, 1965. ICRP Publication 9. Oxford, Pergamon Press.

ICRP TASK GROUP ON THE BIOLOGICAL EFFECTS OF HIGH-ENERGY RADIATION (1966) Radiobiological aspects of the supersonic transport. *Health Physics* **12,** 209–226.

ICRU REPORT 13 (1969) *Neutron Fluence, Neutron Spectra and Kerma.* Washington D.C. I.C.R.U.

ICRU REPORT 16 (1970) *Linear Energy Transfer*. Washington D.C. I.C.R.U.

ICRU REPORT 19 (1971) *Radiation Quantities and Units*. Washington D.C. I.C.R.U.

IVES, P. T., LEVINE, R. and YOST, H. T. (1954) The production of mutations in *Drosophila melanogaster* by the fast neutron radiation of an atomic explosion. *Proc. natn. Acad. Sci., Wash.* **40**, 165–171.

KATO, H., SCHULL, W. J. and NEEL, J. V. (1966) A cohort-type study of survival in the children of parents exposed to the atomic bombings. *Am. J. hum. Genet.* **18**, 339–373.

KAYHART, M. (1956) A comparative study of dose-action curves for visible eye-colour mutations induced by X-rays, thermal neutrons and fast neutrons in *Mormoniella vitripennis*. *Radiat. Res.* **4**, 65–76.

KIMBALL, R. F., GAITHER, N. and WILSON, S. M. (1959) Reduction of mutation by post-irradiation treatment after ultraviolet and various kinds of ionizing radiations. *Radiat. Res.* **10**, 490–497.

KONDO, S. (1965) RBE of fast neutrons to γ-rays for mutations in relation to repair mechanisms. *Jap. J. Genet.* **40**, Suppl., 97–106.

LAMB, M. J., MCSHEEHY, T. W. and PURDOM, C. E. (1967) The relative mutagenic effectiveness of fast neutrons and X-rays in pre- and post-meiotic germ cells of *Drosophila melanogaster*. *Mutat. Res.* **4**, 461–468.

LÉONARD, A. and DEKNUDT, GH. (1967) Relation between the X-ray dose and the rate of chromosome rearrangements in spermatogonia of mice. *Radiat. Res.* **32**, 35–41.

LÉONARD, A. and DEKNUDT, GH. (1968) Chromosome rearrangements after low X-ray doses given to spermatogonia of mice. *Can. J. Genet. Cytol.* **10**, 119–124.

LUNDQVIST, U. (1967) Genetic analysis of eceriferum mutants in barley. *Induced Mutations and Their Utilization*, eds. F. GROBER, F. SCHOLZ, and M. ZACHARIAS, Berlin, Akademie Verlag, pp. 43–45.

LUNDQVIST, U., von WETTSTEIN-KNOWLES, P. and von WETTSTEIN, D. (1968) Induction of *eceriferum* mutants in, barley by ionizing radiations and chemical mutagens. II. *Hereditas* **59**, 473–504.

LÜNING, K. G. and SEARLE, A. G. (1971) Estimates of the genetic risks from ionizing irradiation. *Mutat. Res.* **12** 291–304.

LÜNING, K. G. and SHERIDAN, W. (1964) Dominant effects on productivity in offspring of irradiated mouse populations. *Genetics* **50**, 1043–1052.

LÜNING, K. G. and SHERIDAN, W. (1966) Do recessive lethals have dominant deleterious effects in mice? *Mutat. Res.* **3**, 340–345.

LÜNING, K. G. and SHERIDAN, W. (1968) Dominant effects of recessive lethals in mice. *Hereditas* **59**, 789–797.

LYON, M. F. and MORRIS, T. (1969) Gene and chromosome mutation after large fractionated or unfractionated radiation doses to mouse spermatogonia. *Mutat. Res.* **8**, 191–198.

LYON, M. F., PHILLIPS, R. J. S. and SEARLE, A. G. (1964) The overall rates of dominant and recessive lethal and visible mutation induced by spermatogonial X-irradiation of mice. *Genet. Res.* **5**, 448–467.

LYON, M. F., MORRIS, T., GLENISTER, P. and O'GRADY, S. E. (1970) Induction of translocations in mouse spermatogonia by X-ray doses divided into many small fractions. *Mutat. Res.* **9**, 219–223.

MACHIDA, L. and NAKAO, Y. (1969) Comparison of mutation frequencies induced with neutrons and X-rays in female silkworm pupae. (In Japanese.) 39th Annual Meeting of the Japanese Sericultural Society, Tokyo.

MALLING, H. V. and DE SERRES, F. J. (1967a) Identification of the spectrum of X-ray-induced intragenic alterations at the molecular level in *Neurospora crassa*. *Radiat. Res.* **31**, 637–638.

MALLING, H. V. and DE SERRES, F. J. (1967b) Relation between complementation pattern and genetic alterations in nitrous acid-induced ad-3B mutants of *Neurospora crassa*. *Mutat. Res.* **4**, 425–440.

MANNEY, T. R., BRUSTAD, T. and TOBIAS, C. A. (1963) Effects of glycerol and of anoxia on the radiosensitivity of haploid yeasts to densely ionizing radiation. *Radiat. Res.* **18**, 374–388.

MATSUMURA, S. (1966) Radiation genetics in wheat. IX. Differences in effects of gamma-rays and 14 MeV, fission and fast neutrons from Po-Be. *Radiat. Bot.* **6**, 275–283.

MATSUMURA, S. and MABUCHI, T. (1965) Differences in effects of γ-rays and fast neutrons from Po-Be source on paddy rice. *Seiken Ziho* **17**, 37–39.

MATSUMURA, S., KONDO, S. and MABUCHI, T. (1963) Radiation genetics in wheat. VIII. The RBE of heavy particles from B^{10} (n,α) Li^7 reaction for cytogenetic effects in einkorn wheat. *Radiat. Bot.* **3**, 29–40.

MEDICAL RESEARCH COUNCIL (1956) *The Hazards to Man of Nuclear and Allied Radiations*. Cmd. 9780. London, H.M. Stationery Office.

MICKE, A., SMITH, H. H., WOODLEY, R. G. and MASCHKE, A. (1964a) Relative cytogenetic efficiency of muons and π-mesons in *Zea mays* (L.). *Proc. natn. Acad. Sci. (U.S.A.)* **52**, 219–221.

MICKE, A., SMITH, H. H., WOODLEY, R. G. and MASCHKE, A. (1964b) Relative cytogenetic efficiency of muons and π-mesons in *Zea mays* (L.) and its modification by post-irradiation storage. *Radiat. Res.* **23**, 537.

MICKEY, G. H. (1954) Visible and lethal mutations in *Drosophila*. *Am. Nat.* **88**, 241–255.

MORTIMER, R., BRUSTAD, T. and CORMACK, D. V. (1965) Influence of linear energy transfer and oxygen tension on the effectiveness of ionizing radiations for induction of mutations and lethality in *Saccharomyces cerevisiae*. *Radiat. Res.* **26**, 465–482.

MOUTSCHEN, J. and MOUTSCHEN-DAHMEN, M. (1970) Effects of thermal, monoenergetic and fission neutrons in *Nigella* chromosomes. *Symposium on Neutrons in Radiobiology*, Oak Ridge, Tenn. CONF-691106, 391–402.

MOUTSCHEN, J., MOUTSCHEN-DAHMEN, M., WOODLEY, R. and GILOT, J. (1969) The relative biological effectiveness of different kinds of radiations on chromosome aberrations in *Nigella damascena* seed. *Int. J. Radiat. Biol.* **15**, 525–540.

MULLER, H. J. (1954) The relation of neutron dose to chromosome changes and point mutations in *Drosophila*. I. Translocations. *Am. Nat.* **88,** 437–459.

MULLER, H. J., OSTER, I. and ZIMMERING, S. (1963) Are chronic and acute gamma irradiation equally mutagenic in *Drosophila? Repair from Genetic Radiation Damage*, ed. F. H. Sobels, Oxford, Pergamon, pp. 275–311.

MUNSON, R. J. and BRIDGES, B. A. (1969) Lethal and mutagenic lesions induced by ionizing radiations in *E.coli* and DNA strand breaks. *Biophysik*, **6,** 1–5.

MUNSON, R. J. and BRIDGES, B. A. (1970) Phage mutations induced in bacterial complexes by ionizing particles of different LET's. *Abstr. IV Intern. Congr. Radiat. Res.* 153.

MUNSON, R. J. and BRIDGES, B. A. (1970) Unpublished work.

MUNSON, R. J., NEARY, G. J., BRIDGES, B. A. and PRESTON, R. J. (1967) The sensitivity of *Escherichia coli* to ionizing particles of different LET's. *Int. J. Radiat. Biol.* **13,** 205–224.

MUNSON, R. J., BRIDGES, B. A., NEARY, G. J. and PRESTON, R. J. (1970a) Unpublished data.

MUNSON, R. J., LAW, J. and BRIDGES, B. A. (1970b) Unpublished data.

MURAKAMI, A. (1966) Relative biological effectiveness of 14 MeV neutrons to gamma-rays for inducing mutations in mature sperm of the silkworm. *Jap. J. Genet.* **41,** 17–26.

MURAKAMI, A. (1967) Effect of 5-bromodeoxyuridine (BUDR) on the frequency of 14 MeV fast neutron induced mutations in the gonial cells of the silkworm. *Ann. Rep. Nat. Inst. Genet. (Japan).* **17,** 103–104.

MURAKAMI, A. (1968) Relative biological effectiveness of fast neutrons for the induction of dominant lethals at various stages of male germ cells in the silkworm. *Ann. Rep. Nat. Inst. Genet. (Japan)* **18,** 95–96.

MURAKAMI, A. (1970) A comparison of mutagenicity of 14 MeV fast neutrons on primordial germ cells among five different X-ray sensitive silkworm strains. *Int. J. Radiat. Biol.* **17,** 479–482.

MURAKAMI, A. (1971) Relative biological effectiveness of 14 MeV neutrons for inducing dominant lethals in mature sperm of the silkworm: a comparison of the RBE for dominant lethals and specific-locus mutations. *Jap. J. Genet.* **46,** no. 2, 67–74.

MURAKAMI, A. and ITO, T. (1969) Co-mutagenesis: An interpretation of the effect of post-irradiation treatment with base analogues in the silkworm. *Mutat. Res.* **7,** 479–481.

MURAKAMI, A. and KONDO, S. (1964) Relative biological effectiveness of 14 MeV neutrons to γ-rays for inducing mutations in silkworm gonia. *Jap. J. Genet.* **39,** 102–114.

MURAKAMI, A. and TAZIMA, Y. (1963) Modification of X-ray induced mutation rate in the silkworm by pre- and post-irradiation treatment with halogenated base analogues. *Ann. Rep. Nat. Inst. Genet. (Japan)* **13,** 89–91.

MURAKAMI, A. and TAZIMA, Y. (1965) Relative biological effectiveness of 14 MeV neutrons to gamma-rays in the induction of mutations in germ cells of hibernating silkworm embryos. *Ann. Rep. Nat. Inst. Genet. (Japan)* **15** 120–121.

MURAKAMI, A., KONDO, S. and TAZIMA, Y. (1965) Comparison of fission neutrons and γ-rays in respect to their efficiency in inducing mutations in silkworm gonia. *Jap. J. Genet.* **40,** 113–124.

MURAKAMI, A., KONDO, S. and TAZIMA, T. (1966) Enhancement effect of fractionated irradiation with 14 MeV neutrons on the induction of visible recessive mutations in silkworm gonia. *Ann. Rep. Nat. Inst. Genet. (Japan)* **16,** 109–110.

NAKAI, S. and MORTIMER, R. (1967) Induction of different classes of genetic effects in yeast using heavy ions. *Radiat. Res.* (Suppl. 7), 172–181.

NAKAO, Y. and MACHIDA, I. (1968) The relative biological effectiveness of mutagenic effects induced by neutrons to X-rays in *Drosophila melanogaster*. *Ann. Rep. Inst. Radiol. Sci. (Japan)* 56–58.

NAKAO, Y. and MACHIDA, I. (1970) Comparisons of the RBE of the various genetic changes between X-rays and neutrons in *Drosophila melanogaster*. *Abstr. IV Intern. Congr. Radiat. Res.* 155.

NATARAJAN, A. T. and MARIC, M. M. (1961) The time-intensity factor in dry seed-irradiation. *Radiat. Bot.* **1,** 1–9.

NATIONAL ACADEMY OF SCIENCES—NATIONAL RESEARCH COUNCIL (1956) *The biological effects of atomic radiation* Washington, NAS-NRC.

NEARY, G. J. (1965) Chromosome aberrations and the theory of RBE, I. General considerations. *Int. J. Radiat. Biol.* **9,** 477–502.

NEARY, G. J. (1970) Irradiation of simple biological specimens by charged particles. *The Uses of Cyclotrons in Chemistry, Metallurgy and Biology*, ed. C. B. AMPHLETT, London, Butterworths, pp. 194–203.

NEARY, G. J., PRESTON, R. J. and SAVAGE, J. R. K. (1967) Chromosome aberrations and the theory of RBE, III. Evidence from experiments with soft X rays, and a consideration of the effects of hard X rays. *Int. J. Radiat. Biol.* **12,** 317–345.

NEARY, G. J., SAVAGE, J. R. K. and EVANS, H. J. (1961) The influence of exposure time on the yield of chromosome aberrations in *Tradescantia* pollen grains produced by fast neutrons and gamma-radiation. *Effects of Ionizing Radiations on Seeds*. Vienna, I.A.E.A., pp. 251–257.

NEARY, G. J., SAVAGE, J. R. K., EVANS, H. J. and WHITTLE, J. (1963) Ultimate maximum values of the RBE of fast neutrons and gamma-rays for chromosome aberrations. *Int. J. Radiat. Biol.* **6,** 127–136.

NEEL, J. V. (1963) *Changing Perspectives on the Genetic Effects of Radiation*. Springfield, Ill., Thomas.

NEEL, J. V. and SCHULL, W. J. (1956) *The effect of exposure to the atomic bombs on pregnancy termination in Hiroshima and Nagasaki*. Publ. no. 461, National Academy of Sciences—National Research Council, Washington.

NEEL, J. V. and SCHULL, W. J. (1962) Genetic effects of the atomic bomb (rejoinder to Dr. de Bellefeuille), *Radiol.* **58,** 385–399.

NEUFELD, J., SNYDER, W. S., TURNER, J. E. and WRIGHT, H. (1966) Calculation of radiation dose from protons and neutrons to 400 MeV. *Hlth Physics* 12, 227–237.

NEUFELD, J., SNYDER, W. S., TURNER, J. E. and WRIGHT, H. (1969) Radiation dose from neutrons and protons in the energy range from 400 MeV to 2 GeV. *Hlth Physics* 17, 449–457.

OAKBERG, E. F. and CLARK, E. (1964) Species comparisons of radiation response of the gonads. *Effects of Ionizing Radiation on the Reproductive System*, eds. W. D. Carlson and F. X. Gassner, Oxford, Pergamon, pp. 11–14.

O'BRIEN, K. and McLAUGHLIN, J. E. (1970) Calculation of dose and dose-equivalent rates to man in the atmosphere from galactic cosmic rays. HASL-228 USAEC, New York.

OFTEDAL, P. (1964) Radiosensitivity of *Drosophila* spermatogonia. III. Comparison of acute and protracted irradiation efficiencies in relation to cell killing. *Mutat. Res.* 1, 63–76.

OSTER, I. I. (1963) The mutational spectrum with special reference to the induction of mosaics. *Repair from Genetic Radiation Damage*, ed. F. H. SOBELS, Oxford, Pergamon, pp. 51–62.

PANIKOVSKAYA, L. I. and TROITSKII, N. A. (1968) The genetic effect of intermediate neutrons: deletion frequency and fertility in *Drosophila melanogaster*. *Genetika* 4, 15–20.

PERSSON, G. and HAGBERG, A. (1969) Induced variation in a quantitative character in barley. Morphology and cytogenetics of *erectoides* mutants. *Hereditas* 61, 115–178.

PHILLIPS, R. J. S. (1961) A comparison of mutation induced by acute X and chronic gamma irradiation in mice. *Br. J. Radiol.* 34, 261–264.

POMERANTZEVA, M. D. and RAMAIYA, L. K. (1963) The genetic after effects of the action of fast neutrons on the sexual cells of male mice. *Doklady SSSR* 151, 203–205. (In Russian.)

POMERANTZEVA, M. D. and RAMAIYA, L. K. (1965) The relative biological effectiveness of various types of ionizing radiations. *Trudy Inst. Genet.* 32, 162–176. (In Russian.)

POMERANTZEVA, M. D. and RAMAIYA, L. K. (1969) The mutagenic effect of different types of irradiation on the germ cells of male mice. I. The comparative genetic radiosensitivity of spermatogonia and of the other stages of spermatogenesis. *Genetika* 5, 103–112. (In Russian.)

POWERS, E. L., LYMAN, J. T. and TOBIAS, C. A. (1968) Some effects of accelerated charged particles on bacterial spores. *Int. J. Radiat. Biol.* 14, 313–330.

PURDOM, C. E. (1963) The effect of intensity and fractionation on radiation-induced mutation in *Drosophila*. *Repair from Genetic Radiation Damage*, ed. F. H. SOBELS, Oxford, Pergamon, pp. 219–235.

RANA, R. S. and SWAMINATHAN, M. S. (1967) Relationship between chimeras and mutations induced by ^{60}Co γ rays and 2 MeV fast neutrons at specific loci in bread wheats. *Radiat. Bot.* 7, 543–548.

RANDOLPH, M. L. (1964) Genetic damage as a function of LET. *Ann. N.Y. Acad. Sci.* 114, 85–95.

RBE COMMITTEE (1963) Report of the RBE Committee to the International Commissions on Radiological Protection and on Radiological Units and Measurements. *Hlth Physics* 9, 357–386.

ROSSI, H. H. (1971) Biophysical aspects of high LET irradiation. *Amer. J. Roentg.* 111, 27–30.

RUSSELL, L. B. (1971) Definition of functional units in a small chromosomal segment of the mouse and its use in interpreting the nature of radiation-induced mutations. *Mutat. Res.* 11, 107–123.

RUSSELL, L. B. and RUSSELL, W. L. (1956) The sensitivity of different stages in oogenesis to the radiation induction of dominant lethals and other changes in the mouse. *Progress in Radiobiology*, eds. J. S. MITCHELL, B. E. HOLMES and C. L. SMITH, Edinburgh, Oliver & Boyd pp. 187–192.

RUSSELL, W. L. (1963) The effect of radiation dose rate and fractionation on mutation in mice. *Repair from Genetic Radiation Damage*, ed. F. H. SOBELS, Oxford, Pergamon, pp. 205–218.

RUSSELL, W. L. (1965a) Studies in mammalian radiation genetics. *Nucleonics* 23, 53–62.

RUSSELL, W. L. (1965b) Effects of the interval between irradiation and conception on mutation frequency in female mice. *Proc. natn. Acad. Sci. (U.S.A.)* 54, 1552–1557.

RUSSELL, W. L. (1965c) Evidence from mice concerning the nature of the mutation process. *Genetics Today*, vol. II, ed. S. J. GEERTS, Oxford, Pergamon, pp. 257–264.

RUSSELL, W. L. (1967) Repair mechanisms in radiation mutation induction in the mouse. *Brookhaven Symp. Biol.* 20, 179–189.

RUSSELL, W. L. and KELLY, E. M. (1965) Mutation frequency in female mice exposed to a small X-ray dose at high dose-rate. *Genetics* 52, 471.

RUSSELL W. L. and RUSSELL, L. B. (1959) The genetic and phenotypic characteristics of radiation-induced mutations in mice. *Radiat. Res.* Suppl. 1, 296–305.

RUSSELL, W. L., RUSSELL, L. B. and KELLY, E. M. (1958a) Radiation dose rate and mutation frequency. *Science*, 128, 1546–1550.

RUSSELL, W. L., RUSSELL, L. B. and OAKBERG, E. F. (1958b) Radiation genetics of mammals. *Radiation Biology and Medicine*. ed. W. D. CLAUS, Reading (Mass.), Addison-Wesley, pp. 189–205.

SAVAGE, J. R. K. (1968) Chromatid aberrations induced by 14.1 MeV neutrons in *Vicia faba* root meristem cells. *Neutron Irradiation of Seeds II. I A.E.A. Tech. Rep. Ser.* No. 92, 9–28.

SAVAGE, J. R. K., PRESTON, R. J. and NEARY, G. J. (1968) Chromatid aberrations in *Tradescantia bracteata* and a further test of Revell's hypothesis. *Mutat. Res.* 5, 47–56.

SCHAEFER, H. J. (1968) Public health aspects of galactic radiation exposure at supersonic transport altitudes. *Aerospace Med.* 39, 1298–1303.

SCHAEFER, H. J. (1969) Radiation measurements at supersonic transport altitude with balloon-borne nuclear emulsions. NASA Joint Report NAMI-1068.

SCHULL, W. J. and NEEL, J. V. (1958) Radiation and the sex ratio in man. *Science* **128**, 343–348.

SCHULL, W. J. and NEEL, J. V. (1962) Maternal radiation and mongolism. *Lancet* 537–538.

SCHULL, W. J., NEEL, J. V. and HASHIZUME, A. (1966) Some further observations on the sex ratio among infants born to survivors of the atomic bombings of Hiroshima and Nagasaki. *Am. J. hum. Genet.* **18**, 328–338.

SCOTT, D., SHARPE, H., BATCHELOR, A. L. and EVANS, H. J. (1967) RBE for fast neutrons and dose-rate studies using fast neutron irradiation. *Human Radiation Cytogenetics*, eds. H. J. EVANS, W. M. COURT BROWN and A. S. MCLEAN, Amsterdam, North-Holland Pub. Co., pp. 37–52.

SEARLE, A. G. (1967) Progress in mammalian radiation genetics. *Proc. 3rd Internat. Congr. Radiation Research, Cortina*, ed. G. SILINI, Amsterdam, North-Holland Pub. Co., pp. 469–481.

SEARLE, A. G. (1970) Unpublished information.

SEARLE, A. G. and PHILLIPS, R. J. S. (1964) Genetic effects of neutron irradiation in mice. *Biological Effects of Neutron and Proton Irradiations*, Vol. 1, Vienna, I.A.E.A., pp. 361–370.

SEARLE, A. G. and PHILLIPS, R. J. S. (1968) Genetic insensitivity of the mouse dictyate oocyte to chronic irradiation. *Effects of Radiation on Meiotic Systems*. Vienna, I.A.E.A., pp. 17–25,

SEARLE, A. G. and PHILLIPS, R. J. S. (1971) The mutagenic effectiveness of fast neutrons in male and female mice. *Mutat. Res.* **11**, 97–105.

SEARLE, A. G., EVANS, E. P. and BEECHEY, C. V. (1971) Evidence against a cytogenetically radio-resistant spermatogonial population in male mice. *Mutat. Res.* **12**, 219–220.

SEARLE, A. G., EVANS, E. P. and WEST, B. J. (1969) Studies on the induction of translocations in mouse spermatogonia. II. Effects of fast neutron irradiation. *Mutat. Res.* **7**, 235–240.

SEARLE, A. G., FORD, C. E., EVANS, E. P. and WEST, B. J. (1968) Studies on the induction of translocations in mouse spermatogonia. I. The effect of dose-rate. *Mutat. Res.* **6**, 427–436.

SMITH, H. H. (1962) The reactor as a tool for research in plant sciences and agriculture. *Programming and Utilization of Research Reactors*, I.A.E.A. Symposium, October 1961. New York, Academic Press, pp. 425–438.

SMITH, H. H. (1967) Relative biological effectiveness of different types of ionizing radiations: cytogenetic effects in maize. *Radiat. Res.* Suppl. 7, 190–195.

SMITH, H. H. (1969) Neutron irradiation of seeds as a tool in plant genetics and breeding. *Jap. J. Genet.* **44**, Suppl. 1, 443–453.

SMITH, H. H. and COMBATTI, N. (1967) Factors influencing variation in RBE in irradiation of maize seeds. *Neutron Irradiation of Seeds, I.A.E.A. Tech. Rep. Ser.* No. 76, 26–27.

SMITH, H. H. and ROSSI, H. H. (1966) Energy requirements and relative biological effectiveness for producing a cytogenetic phenomenon in maize by irradiation of seeds with x-rays and monoenergetic neutrons. *Radiat. Res.* **28**, 302–321.

SMITH, H. H., COMBATTI, N. C. and ROSSI, H. H. (1968) Response of seeds to irradiation with X-rays and neutrons over a wide range of doses. *Neutron Irradiation of Seeds II. I.A.E.A. Tech. Rep. Ser.* No. 92, 3–8.

SMITH, H. H., BATEMAN, J. L., QUASTLER, H. and ROSSI, H. H. (1964) RBE of monoenergetic fast neutrons: cytogenetic effects in maize. *Biological Effects of Neutron and Proton Irradiations*, 2. Vienna, I.A.E.A., 233–248.

SMITH, H. H., WOODLEY, R. G., MASCHKE, A. and COMBATTI, N. C. (1965) Relative cytogenetic efficiency of x rays and 28 GeV protons in *Zea mays*. *Radiat. Res.* **25**, 241.

SMITH, J. W. (1971) Personal communication.

SOBELS, F. H. (1967) RBE values for genetic effects of 15 MeV neutrons in relation to stage sensitivity in *Drosophila*. *Int. J. Radiat. Biol.* **13**, 378–379.

SOBELS, F. H. and BROERSE, J. J. (1970) RBE values of 15 MeV neutrons for recessive lethals and translocations in mature spermatozoa and late spermatids of *Drosophila*. *Mutat. Res.* **9**, 395–406.

SPARROW, A. H., UNDERBRINK, A. G. and ROSSI, H. H. (1972) Mutations induced in *Tradescantia* by small doses of x rays and neutrons: Analysis of dose-response curves. *Science* **176**, 916–918.

STEVENSON, G. R. (1964) A personnel dosimetry service for fast neutrons. Radiological Protection Service Report (RPS 373). Radiological Protection Service, Sutton, Surrey, U.K.

STOILOV, M. (1968) Effect of gamma rays and fast neutrons on different maize varieties. *C.r. Acad. Sci. Agric. bulg.* **1**, 221.

TAZIMA, Y. and MURAKAMI, A. (1963) The increase in induced mutation frequency after fractionated irradiation of gonial cells of the silkworm. *Jap. J. Genet.* **38**, 207.

TAZIMA, Y., ONIMURA, K. and FUKASE, Y. (1968) Difference in the proportion of mosaics among mutants induced by 14 MeV neutrons, γ rays and some chemical mutagens in silkworm spermatogenic cells. *Ann. Rep. Nat. Inst. Genet. (Japan)* **18**, 87–88.

TODD, P. (1967) Heavy-ion irradiation of cultured human cells. *Radiat. Res.* Suppl. 7, 196–207.

TRAUT, H. (1963) The linear dose-dependence of radiation-induced translocation frequency in *Drosophila melanogaster* at relatively low X-radiation doses. *Int. J. Radiat. Biol.* **7**, 401–403.

TROITSKII, N. A. and BYLINSKII, A. F. (1966) Relative biological effectiveness of intermediate-energy neutrons (based on phage induction). *Vliyanie Ioniziruyushchikh Izluchenii na Nasledstvennost'*, ed. N. P. DUBININ, Moscow, Izdatel'svo Nauka, pp. 3–7.

TROITSKII, N. A., BYLINSKII, A. F. and FILIPPOVICH, A. S. (1965) Mutagenic efficiency of intermediate neutrons. *Vestsi. Akad. Navuk BSSR, Ser. Biyol. Navuk* **4**, 124–126. (In Russian.)

TROITSKII, N. A., BYLINSKII, A. F. and FILIPPOVICH, A. S. (1966) The mutagenic effect of intermediate neutrons. *Mechanisms of Mutation and Inducing Factors*, ed. Z. LANDA, Prague, Academia, pp. 175–178.

TUTIKAWA, K. (1967) Relative biological effectiveness of 14.1 MeV neutrons in the induction of dominant lethal mutations in the mouse. *Ann. Rep. Nat. Inst. Genet. (Japan)* **17**, 105–106.

UNDERBRINK, A. G., SPARROW, R. C., SPARROW, A. H. and ROSSI, H. H. (1970) RBE of x rays and 0.43 MeV monoenergetic neutrons on somatic mutations and loss of reproductive integrity in *Tradescantia* stamen hairs. *Symp. on Neutrons in Radiobiology, Oak Ridge, Tenn.* Nov. 1969, CONF-691106, 373–390. Also *Radiat. Res.* **44**, 187–203.

UNDERBRINK, A. G., SPARROW, R. C., SPARROW, A. H. and ROSSI, H. H. (1971) Relative biological effectiveness of 0.43 MeV and lower energy neutrons on somatic aberrations and hair length in *Tradescantia* stamen hairs. *Int. J. Radiat. Biol.* **19**, 215–228.

UNITED NATIONS (1958) *Report of the U.N. Scientific Committee on the effects of atomic radiation.* General Assembly document, 13th session, Suppl. No. 17, A/3838. United Nations, New York.

UNITED NATIONS (1966) *Report of the U.N. Scientific Committee on the effects of atomic radiation.* General Assembly document, 21st session, Suppl. No. 14, A/6314. United Nations, New York.

UNITED NATIONS (1969) *Report of the U.N. Scientific Committee on the effects of atomic radiation.* General Assembly document, 24th session, Suppl. No. 13, A/7613. United Nations, New York.

UNITED NATIONS (1972) *Report of the U.N. Scientific Committee on the effects of atomic radiation* (in press).

VAN DER FEER, Y. (1971) Personal communication.

VAN DER FEER, Y., STOUTE, J. R. D. and DE OUDE, A. (1964) De RCN-filmdosismeter. *Atomenergie en haar toepassingen,* **6**, 101–105.

WILHOIT, D. G. and JONES, T. D. (1970) Dose and LET distributions in small-animal sized cylinders for a fission neutron spectrum. *Radiat. Res.* **44**, 263–272.